Science, belief and behaviour

R. B. Braithwaite lecturing at the 1952 Joint Session
of the Aristotelian Society and the Mind Association
(*Radio Times Hulton Picture Library*)

Science, belief and behaviour

ESSAYS IN HONOUR OF
R. B. BRAITHWAITE

Edited by D. H. MELLOR

CAMBRIDGE UNIVERSITY PRESS

Cambridge
London   New York   New Rochelle
Melbourne   Sydney

Published by the Press Syndicate of the University of Cambridge
The Pitt Building, Trumpington Street, Cambridge CB2 1RP
32 East 57th Street, New York, NY 10022, USA
296 Beaconsfield Parade, Middle Park, Melbourne 3206, Australia

First published 1980

Phototypeset in V.I.P. Bembo by
Western Printing Services Ltd, Bristol

Printed and bound in Great Britain
at The Pitman Press, Bath

British Library Cataloguing in Publication Data
Science, belief and behaviour
1. Science – Philosophy – Addresses,
essays, lectures
I. Mellor, David Hugh   II. Braithwaite,
Richard Bevan
501   Q175.3   79-41614
ISBN 0 521 22960 X

Contents

v

Preface

This volume takes the occasion of his eightieth birthday to celebrate
the work and teaching of R. B. Braithwaite FBA, Fellow of King's
College and Knightbridge Professor Emeritus of Moral Philosophy
in the University of Cambridge. All the essays in this collection
have been written for the occasion by his friends, colleagues and
students from Britain and America. The number, at least, of essays
from past and present Cambridge philosophers of science betokens
Braithwaite's successful fostering of the subject here; while the
range of our other contributors will, I hope, dispel any suspicion of
parochial bias. I am especially indebted to Professors Levi and
Schick for their help and encouragement in eliciting the American
contributions.

Braithwaite has made signal contributions to the topics treated in
this volume, and markedly influenced the thought of its contributors. He is perhaps most famous for his great additions to the
already impressive body of Cambridge work on probability and the
philosophy of science. But he has never let himself be cabined there
by the disdainful ignorance of science which continues to impover-
ish much English philosophy. On the contrary, he has maintained
and enhanced a great Cambridge tradition in keeping his metaphysics, ethics and philosophy of mind continually sharpened and
informed by relevant advances in the natural and social sciences,
logic and mathematics.

The manner, as much as the matter, of Braithwaite's philosophiz-
ing is what has attracted the affectionate respect of his colleagues
and friends. By this I mean not just his prose style, but his exhilarat-
ing exuberance in argument. Few philosophers show such cheerful
and unaffected passion for their subject; and none of like achieve-

vii

ment is more free of self-conceit. Braithwaite will found no school, and leave no disciples, as the diversity of styles and opinions in this volume clearly attests: the very idea would only provoke the ribaldry with which he has greeted the philosophical cults of our time. But those of us who revere Braithwaite's irreverence may perhaps be forgiven for trying to reach his own level of pertinent, informed, passionate and disinterested argument. How far we have succeeded in the ensuing pages is not for us to say.

D.H.M.

Cambridge
October 1979

Introduction

The essays in this collection have more in common than the occasion of their production. They all deal, in one way or another, with the impact of science on our beliefs about the world, and how the content and strength of those beliefs do and should affect how we behave towards nature and each other. These have been the central themes of Braithwaite's own philosophy, and, as he has shown, none can be fully understood except in relation to the others. This collection of essays has, therefore, more than a merely conjunctive unity, as I shall now try briefly to exhibit.

The first four essays are about scientific theories. Empiricists like Braithwaite have found difficulties in justifying scientists' criteria for assessing their theories, and Buchdahl argues in the first essay that this is because empiricists exaggerate the importance of inductive criteria. Only by giving more weight to semantic and structural criteria, he argues, can the difficulties, notably the notorious problem of induction, be overcome without lapsing into relativist or subjectivist accounts of theoretical knowledge. Relativism provides the challenge for the second essay also, wherein Jardine looks for a sense in which our successful science can be taken to supply us with absolute knowledge, *i.e.* knowledge whose content is independent of our conceptual and sensory limitations.

Essays three and four deal with the relations of theories to the beliefs on which they are based, and whose truth they serve to explain. Körner relates the explanatory and predictive power of a theory to the way it modifies the organization of its adherents' previous beliefs and concepts. Conversely, Masterman works out, with detailed examples, how T. S. Kuhn's "paradigms", construed as crude ideas associated with a theory's concepts, provide analogies for guiding its development. So construed, she argues, Kuhn's insights amplify and enrich Braithwaite's conception of a scientific theory as a verifiable hypothetico-deductive system rather than supporting subsequent relativist rivals to it.

With Hesse's essay we turn to the question of how a theory supports the laws which follow from it. For Hesse, a theory's function is to articulate inductively relevant similarities between the particular things and events it applies to; and these, she argues, are what license the analogical inferences, from observed to further instances, which laws are used to make. This view of a law, like Braithwaite's, is Humean in taking it to assert (at most) a constant conjunction. My essay, which follows Hesse's, complements hers in this respect by attacking two recently revived and developed anti-Humean positions, which take laws to assert, respectively, necessary truths and contingent relations between universals.

The next three essays all have to do with Braithwaite's accounts of inductive evidence for laws, whether deterministic or statistical, and of statistical probability, *i.e.* chance, and statistical inference. Braithwaite holds no inductive logic: instead, he conceives induction as a self-correcting process that needs only to be reliable in fact, as we think it is; not as something in need of an *a priori* justification which is notoriously hard to supply without begging the question. In the theory of statistical inference, this conception appears also in the Neyman–Pearson criteria for statistical tests, which Braithwaite adopted and explicated in his theory of probability. Levi, in the first of these three essays, shows in detail how much C. S. Peirce's writings on induction prefigured these views; and in the second, Hacking defends Braithwaite's use of the Neyman–Pearson criteria as embodying a sound theory of probable inference.

Kyburg, on the other hand, thinks Braithwaite's is an inadequate theory of probable inference as it stands, and proposes to supplement it in two ways. One is to extend Braithwaite's "Briareus" model, construed as a possible world semantics for chance statements, to cope with irrational values of chance, which occur in statistical theories. The other, more important, is to add a rule of direct inference, suitably circumscribed to avoid the lottery paradox. Kyburg argues that we need this to explain how hypotheses about chances which survive Braithwaite's rejection rules can be used to guide our decision making on particular occasions, without resorting to subjective probabilities or to inductive probabilities conceived as logical relations.

Braithwaite himself recognized this need, but not wishing to eschew subjective probabilities (*i.e.* degrees of belief with a probabilistic measure) as Kyburg does, he put it differently. Since his theory tries to make chance do the work of inductive probability, he has to say how and why knowing the chance of an event should affect behaviour whose outcome depends on whether the event occurs. He tackled the problem by explaining why this knowledge should fix one's degree of belief in the event's occurrence. Now a Bayesian who admits chance at all will naturally suppose that this is just how knowledge of it should affect behaviour, *via* fixing degrees of belief which combine with utilities to determine preferences as prescribed by Bayesian decision theory. But recent literature, *e.g.* on Newcomb's paradox, shows that rational preference also depends on whether one's actions help to cause events they make probable, or are merely evidence for their occurrence. Jeffrey, in the essay which follows Kyburg's, illustrates and discusses the problem this poses for decision theory, though without committing himself to any of the possible solutions he flirts with.

Jeffrey's essay shows how even Bayesian decision theory may need to take chance as more than objectified degree of belief. In the next essay, Harsanyi likewise distinguishes chance from subjective probability; but he is concerned here only with the latter, which he invokes to tackle two problems in the theory of games. In the first, he uses Bayesian decision theory to analyse the games in which, as in real life, players have less than complete information about other players' preferences, strategies – and about the incompleteness of the others' information. In the second, he shows how generally to apply the Bayesian apparatus of one-person decision theory to the study of *n*-person noncooperative games.

It is obvious that games theory involves ethics, since it treats of people's dealings with each other. The converse has not always been so obvious, and Braithwaite's inaugural lecture on the use of games theory in ethics was a considerable innovation, as Schick remarks in the concluding essay. In it, he deals with how the distribution of goods in a population should be related to people's wants, rights and resources; and in particular, how far the liberal's insistence on justice can be pressed when it conflicts with the claims of welfare.

Even from these brief introductory remarks it will be evident that the essays which follow tackle a wide range of topics, for all their interrelatedness; and their collective scope is in itself a tribute to the burgeoning of Braithwaite's ideas. If, between them, these essays prove as fruitful as Braithwaite's work has been, the authors will be more than pleased; and it is in this hope that we celebrate his eightieth birthday by placing the essays before the philosophical community.

D.H.M.

1 Neo-transcendental approaches towards scientific theory appraisal

GERD BUCHDAHL

In surveying the historical evidence concerning the growth of science, we encounter broadly three groups of considerations or criteria which appear to have determined standards of acceptability, appraisal and, indeed, of actual lines of research. An explicit reference to such a triadic scheme of considerations (to my knowledge) was made for the first time in Kant's *Logic*. Whilst the inductive process, starting with empirical data, yields certain – occasionally 'lawlike' – generalizations whose "probability"[1] it thus determines, certain additional criteria must be satisfied, responsible (according to Kant) for the "possibility" and systematic "unity" of a scientific theory and its fundamental explanatory concepts (Kant 1974: 92–3; *cf*. A770/B798). Since Kant's time there have been a number of attempts to develop such methodological approaches under a variety of labels, such as in the writings of Whewell (1967: Part II, Bk. XI), Schleiden (1849–50; *cf*. Buchdahl 1973) and Hertz (1956: 1–14) during the nineteenth century, and during more recent times by writers like Meyerson (1930), Burtt (1950), Holton (1973: I), Harré (1964), *etc.*, as well as myself (*cf*. Buchdahl 1970: 226–7).

In this essay I want to explore the rationale of such a structure in terms of a neo-transcendentalist approach; 'neo', because my account involves a fresh amalgamation of certain central themes occurring in Kant as well as Husserl, but treated there with insufficient explicitness.

---

[1] I use double quotes in citing from an author's own terminology or text, especially where this has a somewhat technical meaning, differing from the usual. For all other purposes, I use single quotes. [The other essays in the volume use single quotes when mentioning words, expressions or sentences (or their meanings), and double quotes for other purposes – Editor.]

1

1   The prime model to which in Kant's eyes such a scheme was immediately relevant was Newton's *Principia*; in particular, its central concept of universal attraction. Kant's response to this system embodies in outline the major features of his philosophy of science. He is quite explicit that in the light of the explanatory and predictive successes of *Principia* it seemed almost impossible for him and his contemporaries to "think up" some alternative system of physical explanation (*cf*. Kant 1953: 83). Yet at the same time, Kant insists, it is not enough to rest with the inductive or hypothetico-deductive [h-d] support for the hypothesis of universal gravitation. It is necessary, in addition, to show that such a hypothesis constitutes a genuine possibility ("real", of course, and not just "logical"). As outlined in his *Metaphysical Foundations of Natural Science*, that implied the need for a "construction" of a viable conception of "matter", *via* an "empirical analysis", in such a way that attraction-at-a-distance emerges as a 'natural' consequence. Thus, if we consider what contemporary physics *means* by matter, it will be found to connote the idea of filling space; and this, in turn, is a notion that involves 'necessarily', as it were, reference to repulsive and attractive forces (Kant 1968: 17–22; 1970: 6–9; *cf*. Buchdahl 1971:36–7). Evidently, this analysis had the aim of removing the air of paradox then surrounding the concept of 'action at a distance'.

It will be noticed that this analysis is "empirical" in availing itself of meanings that are broadly entrenched in contemporary Newtonian physics, and implicit in its growth between the time of Newton and Kant. It is not, of course, 'empirical' in the sense of appealing to further *specific* observations or experiments but instead focuses on (what Whewell called) conceptual explication; it is in *this* sense that Kant labels these accounts "metaphysical", rather than physical. The *Metaphysical Foundations* seeks to formulate a number of such explications, *e.g.* of the possibility of applying joint velocity measures to matter in motion, of formulating certain laws of conservation such as of mass, velocity, momentum, *etc.*

'Conceptual explication' is a more sophisticated version or re-reading of older attempts to place science on a 'metaphysical basis', in the traditional sense of this term, as understood in the writings of, say, the major seventeenth and eighteenth century scientists and philosophers of science, *e.g.* Descartes and Leibniz, Fermat, Euler, Maupertuis – to name just a few. Moreover, 'conceptual explica-

tions' have reappeared and, indeed, during the last twenty years
have usurped the centre of the stage in the philosophy of science.
Whilst Whewell had made this theme one of the main tenets of his
account of the philosophy of theory construction, it subsequently
surfaced again in J. B. Conant's influential reconstruction of the
history of science in terms of a succession of "conceptual schemes"
(1947); a notion which was taken up by his pupil T. S. Kuhn (1962),
whose "paradigm"-historiography is too well known to require
here any discussion. In a similar vein, Gerald Holton (1973) and
Rom Harré (1964) have emphasized the conceptual or "thematic"
dimensions of science. Finally, the notion of a "metaphysical hard
core" of science, in the work of Imre Lakatos (1970), is clearly a
latter-day descendant of 'conceptual explication', viewed as one of
the components – I will call it the 'Semantic Component' – of
theory construction and appraisal.

So much for "possibility"; let us now turn to "unity". The
'systemic' aspect of science as such has already been alluded to.
Systematization, though frequently treated in isolation, more often
has also involved certain methodological ideas and maxims; simplicity, continuity, economy; analogy, conservation, symmetry,
'finality' are just some that have at different times and places determined the particular form taken on by systemic articulation. However, the function of such ideas and maxims is not merely confined
to guiding the direction of scientific research and discovery; historically they have been treated commonly as contributing to the
special explanatory force of a theory. In other words, they act as
(part) criteria for the acceptability of a theory. Moreover, as in the
case of the semantic component, they originate as expressions of the
metaphysical basis of science. Thus, continuity and finality (*i.e.*
teleological explanations) are in Leibniz declared to possess an
"architectonic" status: edicts of, and grounded in, the divine
architect. In Whewell, "consilience of inductions" as well as "sim-
plification of theories" are looked on as basic criteria for the acceptability of theories; as is the idea of "economy" in the philosophical
scheme of Ernst Mach. Systemic articulation, when further determined by such methodological maxims, thus evidently forms an
additional component of theory construction; I will call it 'the
architectonic component', with Kant's use of this term, meaning
"the art of constructing systems" under the guidance of "ideas"
(A832/B860).

Figure 1 exhibits in summary fashion such a triadic structure of components of theory construction and appraisal. (The joining of 'construction' and 'appraisal' is deliberate, as implying that no sharp distinction is drawn between these two sides of scientific method.) Kantian in inception, it seeks to incorporate subsequent developments and insights, in so far as they exhibit affinity with the original source. In particular, the schema emphasizes the strong interaction that has historically existed between the physical or inductive component, and the two non-physical components (semantic and architectonic). Thus, the inductive and systemic articulation of

Figure 1 Methodological components of scientific theory appraisal (A Kantian formulation)

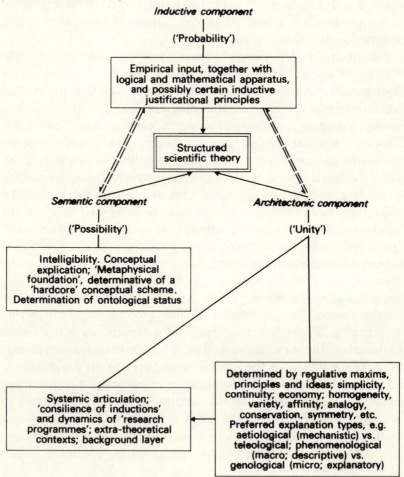

Newton's *Principia* was to a considerable extent instrumental in determining the details of the conceptual explication of the Kantian notion of "matter" (already mentioned) that emerged under the guise of various 'dynamic' and 'field' theories during the late eighteenth and nineteenth centuries. Or to mention an example of influence from the regulative on the inductive component: the conservation idea was obviously to a considerable extent responsible for the direction taken by nineteenth-century physics. But in turn, the theoretical (inductive and systemic) success of conservation principles during that period gradually led to the elevation of conservation to the status of a 'justificational' idea; witness the writings of Poincaré and Meyerson, in which conservation principles no longer play the part merely of physical hypotheses that might be held to 'explain' those 'empirical generalisations' or 'laws' deducible from them – to use the Braithwaitean formulation (1955). On the contrary, the notion of 'explanation' receives now a different twist. The age-old metaphysical idea of the '*ex nihilo nihil fit*', and its various later equivalents involving 'conservation' are felt to bestow a special force, and a much deeper sense, of explanation on the hypotheses involving this idea.

2 We cannot discuss in any systematic fashion the methodological scheme presented here, since we shall be primarily concerned with the rationale and logical status of such a structure rather than its formal details, or the various uses and interpretations which it has received in the course of history. For this essay, it is sufficient to note a few general aspects. As already seen, the three components mutually interact with one another. Furthermore, each must be supposed to yield some *prima facie* justificative force towards acceptability, subject to that generated by the other two. But there are no fixed rules governing the individual 'confirmation-weights' derived from these components, nor the method of compounding these weights. In our historical example of gravitational attraction, for instance, Newton's conceptual explication of matter was such that it necessarily militated against the support derived from the other two components, since the action of matter at a distance – 'where it wasn't' – was for him an 'irrational idea' (*cf.* Buchdahl 1970: 215). But whilst in the eyes of continental scientists (Leibniz, Huygens and others) – who were operating with a similar conception – this altogether nullified any supposed 'strength' that might flow from the inductive and systemic aspects, leading them to reject Newtonian gravitational theory altogether, Newton chose to ignore the conceptual 'anomaly', or sought to placate it in some way or other. And in the sequel, the inductive prestige of the theory gradually forced the replacement of this conceptual 'anomaly' by an explication more consistent with inductive 'evidence'. In the words of Stallo, 'If we reverse the proposition that a body acts where it is, and say that a body is where it acts, the inconceivability disappears at once' (1885: 145).

More generally, different periods in the history of science, and, in consequence, various philosophical schools, in their reflections on the nature of scientific theory, have not only given different accounts of the logical significance of these components but have also placed different emphases on their respective relevance and importance. Traditionally, seventeenth- and eighteenth-century 'rationalist' schools of thought put the main stress on what we can now regard as the semantic component. Thus, Descartes' reputation (somewhat unjustly exaggerated in most accounts, owing to lack of understanding of the complexity of the methodological situation) for an '*a priori*' approach in science is due to his focusing on the problem of explication of the basic concepts of dynamics. Explication here took the form of enunciating certain 'rules' or 'laws of nature' ('conservation of motion', law of inertia) primarily based on ideas of constancy and causation, supposedly grounded in God. It was thus that one was supposed to possess an independent foundation for a theory of dynamics. This was, of course, a way of saying that the basic concepts of a science need to be formulated in a general way before one can get started on empirical description and hypothesis formation, and that to this end one must begin by relating such concepts (as special cases) to very pervasive aspects of all human experience; in Descartes' case: causation. (Kant will later use the same method in connection with his "proof" of the law of inertia, which he treats as a special case of the law of causation, previously 'shown' to be a necessary ingredient in his definition of *all* experience of change in general, 'scientific' or otherwise. See Buchdahl 1969: 674–8.) By contrast, 'empiricist' philosophers originally concentrated almost exclusively upon the 'inductive component' (Locke, Hume), though they gradually widened the net so as to include the aspect of systemic articulation, illustrated by the succession of empiricist writers, from Berkeley, through Mill to Braithwaite, Nagel and many others in our own century.

Braithwaite's was here one of the classical presentations of this tradition. The inductive component consists of the empirical input (individual observational data) processed by means of an inductive apparatus to yield generalizations that form the empirical basis of some given scientific theory. The theory itself is represented as an h-d hierarchical structure (the 'systemic articulation'). The most noteworthy aspect is that many of the other criterial features (part of the semantic and architectonic components), are almost completely reduced to the inductive and systemic side of theories. Thus, the existence of analogy (the force of 'models' for a theory) is seen in purely formal terms, *via* the identity of deductive structure that may be held to exist between a theory and its model. Particular explanation types, such as teleological explanations, have no independent significance but are similarly reduced to purely formal aspects of the h-d system. The 'lawlikeness' of certain empirical generalizations, again, is accounted for in terms of formalist "frillings" which interprets their lawlikeness as a function of their positions within a "deductive system" (Braithwaite 1955: 317 and 317n); whereas, historically, lawlikeness had been explained in terms of the edicts of a divine lawgiver, or its modern equivalent: an irreducible property of nomic necessity; or – as in the sophisticated theological version of Kant's *Critique* – *via* a teleologically organized system ultimately involving the *analogue* of a divine ground (*cf.* A681/B709).

The semantic aspects are perhaps the most interesting in this account. There is the sharp distinction between the theoretical and empirical; the latter (as explicitly stated in Braithwaite's discussion of "general statements") reduced to particular statements "expressing observable facts about particular observable things" (Braithwaite 1955: 84). Finally, there are statements involving theoretical terms not reducible to corresponding observation statements. Such statements may function as basic 'explanatory' hypotheses; *e.g.* laws of conservation, the law of inertia, the second law of thermodynamics, *etc.* Now where the rationalists registered the logic of such statements and their corresponding concepts by means of the sort of 'explications' just discussed, Braithwaite accounts for their logical status once more simply in terms of their formal position within the h-d system, showing that the basic terms involved in such hypotheses are "Campbellian" in character, *i.e.* lacking in 'correspondence rules' (Braithwaite 1955: 77, 99–144).

It is not relevant here to go further into the details of this ingenious system. I am more concerned with its philosophical assumptions. It is evident that its entire centre of gravity is located in the individual observation statements (couched in an independent observation language), and dependent on the inductive assumptions that would allow us to formulate ('infer') corresponding empirical generalizations; plus, finally, the systemic articulation of these generalizations in deductive form. Faithful to its empiricist ancestry and spirit, no room is left for the remaining components of the triadic structure elucidated above. Whilst no doubt conceptual explication and regulative determination (in terms of the classifications mentioned in Figure 1 above) might be acceptable as purely psychological or heuristic or methodological factors involved in the *construction* of theories, they have no significance for their rational appraisal.

3   Now a first objection to this empiricist impoverishment of the methodology of appraisal is that it runs counter to the judgement of the science of past and present. *Per contra*, it is precisely the virtue claimed by our triadic structure that it mirrors judgements of some of the greatest among scientists and philosophers of the past concerning the acceptability of scientific theories, being thus no arbitrary scheme but one that has been found conventionally to define the very nature of science. (For a special illustration of this claim, in relation to the scientific work of Kepler, see Buchdahl 1972.)

Against this, it will no doubt be urged by the empiricist that history is no judge of the rationality of theory justification. This is a dispute that can obviously not be resolved within the compass of this essay in any general terms. My strategy will therefore be confined to the following. I shall suggest that the empiricist's focusing on the inductive component is itself the result of an over-optimistic assessment of its strength as a rational criterion. Furthermore, I shall claim that the fatal weakness, and indeed lack of intelligibility, of the empiricist interpretation of this criterion can be characterized most succinctly by the criticism that the empiricist's attitude to 'empirical fact' is tantamount to treating it as though it possessed the status of a Kantian 'thing in itself', involving a position of what Putnam has recently characterized as "metaphysical realism". The alternative position, Putnam's "internal realism" (identical evidently with Kant's "empirical realism", suspended within the boundary of "transcendental idealism") will then be used to sketch a transcendental account of the status of the rational criteria of scientific appraisal, and which may be viewed as an alternative to the now common subjectivist and relativist responses to the difficulties of the empiricist viewpoint.

Consider, then, first some of the difficulties surrounding the empiricist's philosophical attitude towards the inductive component. To start with, contrary to what seems to be taken for granted, there is the now well-known absence of a sharp dividing line between observational and theoretical language, and the connected phenomenon of 'meaning-variance', which already shows that the inductive component's viability depends on that of the other two, and cannot thus be appraised in isolation. Furthermore, the notion of the 'empirical input' is usually understood by the empiricist in terms of 'bare observational data'. Yet it is difficult to make philosophical sense of the implied assumption that such data can be peeled off the skein of scientific discourse – a difficulty frequently discussed under the heading of "the myth of the given". (See for instance, Sellars 1963: 161.) The opposite doctrine is best couched again in a Kantian garb: sensation cannot be severed from thought, or "matter" from "form", or a bare "given" from what displays itself, if at all, only in a conceptual wrapping.

In more general terms, it may perhaps be asked what relation our various methodological components – especially the non-physical ones – bear to the problem of 'scientific truth'? The answer will be as follows. It is only inside the framework that a systematic research programme can first get off the ground. The framework will formulate concepts, question-areas or domains of research inquiries, *etc.*, *etc.* Nor is it possible to say that within the terminological constraints thus generated, akin to a Wittgensteinian 'network', experiential data could yield yes-or-no answers to questions which the scientist "has himself formulated" (Bxiii). For if, as Kant himself recognizes in another place, 'the principles according to which we perform experiments must themselves first be taken from our knowledge of nature, and hence from theory itself' (Kant 1965: 6), then there will be (as already noted) no hard and fast division between the methodological framework, the theoretical network and the 'facts' supposedly explained by these agencies. In which case, the problem of scientific truth recedes rather beyond the horizon.

This position has been revived recently in Hilary Putnam's recantation of his older "metaphysical realism", in favour of an "internal" realism (Putnam 1978: 130, 135–6). We cannot make sense, he writes, of a notion of 'THE WORLD [which] is supposed to be *independent* of any particular representation we have of it . . . [We] may retain THE WORLD but at the price of giving up any intelligible notion of *how* THE WORLD is'; whereas, he argues, the world is in fact always "theory-relative" (1978: 125, 132). He does not specify in detail the sort of 'theories' he has in mind: are they specific scientific theories, or Kuhnian "paradigms", or some methodological structure or other, governing the theory of the world we happen to be studying; or some common 'language'; or is the intention simply to insist in general terms that the notion of "a world" is 'condition-relative' as such, with the particular conditions involved depending on some chosen philosophical or metaphysical scheme?

Rather than interpreting Putnam's somewhat sketchy account, I shall use his hints and link them to a historical tradition involved in such a position by expressing it in terms of something like a Kantian categorial framework; or rather – in line with the main theme of this essay – of a set of methodological ideas, principles and maxims, on the lines discussed in Kant's chapter on the "regulative employment of reason" in the first *Critique* (A642/B670ff), and similarly in the Introduction to the *Critique of Judgment*. (For this, see also Buchdahl 1969: 495–530.) Whatever the details of Putnam's own position, it evidently represents (what he describes as) "a soft (and demythologized) Kantianism" (Putnam 1978: 138). By contrast, for the 'metaphysical realist' "THE WORLD ends up as a Kantian 'noumenal' world, a *mere* 'thing in itself'" (1978: 133).

No doubt, empiricists like Braithwaite may be surprised to find themselves labelled "metaphysical realists" in the sense of believing in 'things-in-themselves'. But such a response can already be found in Kant who argued that, of all people, "*Hume* took the objects of experience as things in themselves" $[T_m]$, since they univocally "filled" the whole of logical "space", leaving no room for 'alternatives' (Kant 1956: 50, 54); not as in Kant's own philosophy, where objects can be regarded from two different points of view, as "appearances" $[T_a]$ or as "things in themselves" $[T_n]$, in a manner to be explained presently (*cf.* Bxix.n). Thus Hume's 'objects', though being received passively into the mind, *in their nature* owe nothing to the cognitive agent: 'For since all actions and sensations of the mind are known to us by consciousness, they must necessarily appear in every particular what they are, and be what they appear' (Hume 1946: 190; *cf.* Buchdahl 1969: 356). (And let us note, for the sake of what is to follow, that there is no apparatus here by means of which as such to transcend these 'moments of presentational immediacy', except *via* psychology (Hume), or inductive or deductive logic, as in Braithwaite and other like-minded philosophers.)

Not surprisingly, this philosophy has usually found it difficult to come to terms with the so-called 'scandal of induction' – at least in its 'old' form. Clearly, 'assumptions' concerning 'uniformity' and similar properties of nature have usually been intended to involve not just a 'formal' but, rather, a 'substantial' claim concerning what 'is the case', even though *that* be *in principle* unknowable. Hence, for an empiricist metaphysics, not only the level of the concrete and particular but also that of the general will ultimately have to possess the character of a 'thing in itself'.

The epistemological failure of this position has been noted in a number of philosophical traditions, including the Kantian. The 'metaphysical' claims just noted, it is argued, *must* fail, for since the world, as 'given', according to that doctrine, is dissociated as such from any 'subjective' conceptual functional conditions, there can be no cognitive access (even in general terms) to what lies beyond the supposed concrete particular 'data'. So unless the world beyond these data could in some sense 'reveal itself', *via* what Leibniz and Kant call "intellectual intuition", the empiricist must end up a sceptic, in the well-known fashion of, say, a Locke, a Hume or a Russell. By contrast, the "internal realist" (a transcendental idealist in disguise) will deny that the notion of a 'world in itself' makes any sense, and will thus avoid the problem. This does not mean that an internal realist or transcendentalist would deny that for science 'there is a real world to be discovered'. But he will insist on the proviso that any notion of 'the world', gradually emerging in the context of scientific discovery, is one in which 'material' and 'formal' (conceptual) aspects are inextricably intertwined.

4 Such a position has some interesting consequences for any appraisal of the status of the methodological framework previously elucidated. First, we can now see that any special priority given to the inductive component, in so far as this rested on some implicit and (usually) unrecognized metaphysical assumptions involving the positing of a 'world in itself' is no longer tenable. In so far as its status depends on its functioning as a methodological condition of the pursuit and justification of scientific theorizing, thereby rendered "really possible" (to use Kant's transcendental jargon), the inductive component's associated "validity" is contingent upon exactly the same considerations as support any other methodological vectors, such as the semantic and architectonic in our scheme: a result which is in line with the Quinean argument that it is not so much theory depending on induction but rather induction depending on the achievements of scientific theory (Quine 1969: 135).

Secondly, the replacement of the metaphysical approach by a transcendental position supplies us with the means of confronting afresh some of the recent relativist and idealist responses (*e.g.* the appraisals of Kuhn and Feyerabend) to the empiricist debâcle. Putting it in somewhat crude and over-simplified fashion, their argument goes that since methodological prescriptions do not yield certainty of truth, they must logically and ontologically be idling.

However, it seems that more often than not the anti-methodologist's response has taken the form of a reaction against a *particular* assumption, *viz.* the inductivist–empiricist *metaphysic*; a reaction, moreover – as frequently happens – which proceeds by unconsciously swallowing its opponents' metaphysical assumptions. Briefly, its critique ultimately reduces to the complaint that since an *absolute and final* truth about a world–in–itself is not obtainable, then 'anything goes'. (Alternatively, and for precisely the same reasons, some anti-*epistemological*, purely heuristic, position, such as L. Laudan's (1977) 'problem-solving' solution is adopted.) A less destructive approach might however suggest instead that the definition of 'the world', or 'the world studied by science, with its received methods', is still up for grabs. Being as yet 'undefined', a preliminary task should be to effect a 'constitutive account', in terms of a framework that will first generate the very concept of such a world.

This is meant to forestall the complaint that if there is no *metaphysically-founded* methodology, then we have no grounds for adopting *any*. On the contrary, on the transcendentalist approach to be spelled out presently, those methodologies can now be taken seriously that have so far yielded the most powerful theoretical insights and practical applications; as supplying the only rationale for adoption. Moreover, we shall in principle assume – in the light of our historical understanding – that methodology will develop dynamically with time. There is no need to adhere slavishly to received schemes, if novel approaches force themselves on our practice.

5 Now it is not, of course, the case that 'history' by itself and *as such* sanctions or validates the adoption of any methodology that may in fact have been adopted by science during the past and present. For such a validation presupposes a transcendental position or argument which interprets the concept of 'the world' in a novel way. (For a special development of this, *cf.* also Habermas 1971: 194 and *passim*.) I shall try and explain.

The verbal account of a transcendental argument usually takes the form of positing certain conditions subject to which alone something or other (modes of speech, kinds of objects, types of experience, *etc.*, *etc.*) is rendered "really possible". A great deal of ink has been spilled over the question whether for any given kind of thing thus 'made possible', one and only one unique set of conditions has to be, or is, capable of being, posited. I shall take the view that the basic structure of the transcendental approach is unaffected by an answer to this problem. Kant, for instance, posited human sensibility (including its formal side, space and time) and the categories of the understanding to be conditions for the possibility of empirical objective reality, with the implied suggestion that these conditions were both necessary and unique for the constitution of nature, taken as the aggregate of things. On the other hand, at a level 'higher up', *viz*. that of "the *order of* nature", though using a similar approach, here towards scientific theory construction, by postulating certain methodological principles or maxims (*cf.* the 'regulative maxims' of our schema), he is not at all insisting on uniqueness as a condition for the possibility of the idea of a 'system of the world' (or order of nature) and of the corresponding scientific theories for such a system. In any case, the maxims were clearly derived from an empirical survey of the history of scientific thought, and of the methods it employed, and their specific forms were thus altogether contingent.

A more difficult problem is posed by the notion of 'making something possible'. Traditionally, this mode of speech occurs in the formulation: '*how* is such and such possible?' The implication here is that the 'object', the 'possibility' of which is to be accounted for, is in some sense 'actual' – which is certainly true in the case of 'the world', though less obviously so in the case of 'the order of nature', of 'causal systems' and the like; whilst the problem of the 'how' is evidently due to certain basic assumptions that seem incompatible with the object's 'actuality'; unless, that is, certain conditions are first provided. For instance, Kant took it for granted that we have experiential knowledge of sequences of states of a thing but argues that this is only possible if individual perceptions, which occur *as such* in a quite "accidental order", are in some sense connected necessarily, or in some rule-like fashion. (*Cf.* the argument of the Second Analogy.) At the next level up (that of theoretical science) the argument recurs: the whole idea of a systematic interconnection between the objects of our world can acquire empirical significance only on condition that certain methodological notions may be presupposed.

Using insights provided by Husserl, especially as enunciated in his *Idea of Phenomenology* of 1907, I want to give the transcendental approach a sharper and somewhat more radical formulation, in order to bypass this confusing relationship between the 'actual' and the 'possible'; indeed, in order to avoid the awkward intrusion of 'actuality' altogether. At the start, in formulating the essentials of his phenomenologist–transcendentalist method, Husserl, in connection with the problem of the possibility of knowledge of the transcendent, writes: 'If I do not understand *how* it is possible that cognition reach something transcendent then I also do not know *whether* it is possible' (1964: 29). And he goes on to suggest that any intrusion of a 'seemingly' *actual* level of reality must be strictly avoided; the focus must be entirely centred on possibility. His method consists in so-to-speak 'putting actuality on ice', thus (we may say) "uncoupling" the 'phenomenology' of any situation from its 'ontology', by a process of "bracketing", sometimes also called "reduction"; which for him usually means a relating of the 'phenomenon' to 'consciousness' as its object, or to a transcendental ego.

I want to modify as well as to radicalize this notion of reduction by bringing it more into accord with Kant's basic intentions; which

Husserl thought *he* was clarifying and developing! (A nice example of the hermeneutic circle!) Let me therefore define the transcendental approach as the doctrine that any object, and in particular, any knowledge of an object, is 'really possible' only on certain conditions. Furthermore, in cases where abstraction is made from these conditions, or where these are assumed unavailable, or altogether suppressed, I shall say that the object has been subjected to a 'transcendental reduction'; and that in such an event it is to be regarded as an 'object' only in a "transcendental sense" – as what Kant calls "a transcendental object" $[O_t]$ for short. (For some classical passages in Kant explaining this particular use of the term, *cf*. A251; A247/B304; A266/B322.)

Going in the reverse direction: where certain conditions are satisfied, *e.g.* where (as in the Kantian case) a context of "sensibility" and "understanding" is provided, yielding the "real possibility" of an "empirical object" (*quâ* "appearance" $[T_a]$), we shall say that a 'realization of the object' has been effected. (For "realization", *cf*. A146/B186.) Finally, following Kant's usage, we shall say that any conditions which generate 'the real possibility' of an object, or of a system of objects, simultaneously thereby obtain "objective validity" (*cf*. A89/B122); which is what Kant means by providing for them a "transcendental deduction". (*Cf*. A671/B699 for the methodological case.)

Such a reduction–realization technique may be said to define the 'philosophical grammar' that bounds possible objects; *e.g.* (again choosing Kantian examples), it explains the sense in which the world can be said to be wholly deterministic, yet compatible with moral freedom; in which it is finite or infinite; in which it exhibits systemicity, *etc*. The technique gives concrete expression to the way in which transcendental accounts are *about* experience and its objects (their 'possibility' in general), compared with those contexts where we are dealing with questions arising *within* experience (*cf*. Brittan 1978: 11). It clearly defines the sense in which such accounts are '*a priori*', with the reduction–realization process as the functional model for this *a priori*.

Space forbids discussion of the details of this transcendental structure and of the powerful means which it affords for an understanding of transcendental philosophy. I will only note one point, concerning Putnam's "theory-relative world". Talk of a 'world' in abstraction from 'theory' (*i.e.* transcendental conditions), is equivalent, according to Putnam, to treating it as a 'thing in itself'. However, it may now be seen that this can be understood in three quite different senses. (1) This defines the 'thing-in-itself' $[T_m]$ as an object assumed not to require the notion of transcendental conditions at all, and hence no reduction–realization procedure either. It is in this sense that Kant labelled Hume's and Leibniz's 'objects' "things-in-themselves". (2) The second sense defines the thing-in-itself as a 'transcendental object' $[O_t]$, *i.e.* one which has not yet been subjected to a process of realization, or where that process has failed to achieve its objective. (3) The third alternative $[T_n]$ (according to Kant) is one where the object is viewed as a realization of O_t that would result if humans were capable of "intellectual intuition" – a rendering of a certain Leibnizian doctrine which Kant of course rejected. Hence T_n (the "noumenon") has no epistemological (and for Kant, only an ethicological) significance, since man is only capable of "sensory" and not "intellectual" intuition. We thus get the result that whilst an epistemologically conceived T_n is radically *inconsistent* with a Kantian position, T_m is not so much inconsistent as the expression of an alternative philosophical doctrine, *viz.* "metaphysical realism"; whilst O_t, for the latter, would simply be a meaningless concept. I suspect that Putnam has availed himself of the logical impossibility of a "positive" interpretation of T_n within his own 'Kantian' position, to give the (I think misleading) impression that the latter 'refutes' the 'realist' T_m.

All this is intended to show that the transcendentalist way for which we have been arguing here must not – as most philosophical doctrines should not – be viewed as a 'refutation' of possible alternatives, but as a proposal to look at things in a certain way. So all one can claim on behalf of the transcendentalist alternative is that it offers a heuristically more powerful tool for interpreting, or doing justice to the sentiments of, the history of methodological reflections on science.

6   Let us now return to the transcendentalist treatment of the status of methodological frameworks. We have seen that a 'realization' of O_t under the conditions of space, time and categories yields the aggregate of empirical objects which, taken together, constitute (what Kant calls) "nature" (Kant 1953: 54). In a parallel way – and this is the case we are interested in here – certain methodological maxims and principles serve as conditions for the 'realization' called "the order of nature" (Kant 1951: 21, 23), identified as the systematic interconnection of things, with the aid of certain 'theoretical concepts' – Kant's examples are earths, water, air, salts, phlogiston, *etc.* – whose status Kant describes on lines of Braithwaite's "Campbellian" terms (A646/B674; Bxiii).

Of central importance here is Kant's contention that such methodological notions have more than merely "logical" force; but in addition they must possess "transcendental" status (A650/B678; A645/B682). The significance of this seldom well-understood contention becomes plain in the light of our reduction–realization account. Regulative principles and maxims would have no more than methodological, not to say psychological, import if we supposed them to apply to a world possessing the status of a 'thing in itself' (see *e.g.*, A679/B707). *Per contra*, if we define 'the world' as subject to transcendental conditions, with a reduction–realization 'history' as an essential part, then such a world (here in the sense of the system or order of nature) has to be construed as the resultant of a process of 'realization' which, operating upon nature regarded as a transcendental object $[O_t]$, proceeds *via* the injection of the principles given by some methodological schema or other.

So, relative to a world taken in the sense of "appearance" $[T_a]$, *i.e.* in so far as its objects "are given to us" (A654/B682), our methodological framework obtains an autonomous function. Though its specific details may be due to historical conditions, its status has changed: from being derived *from* that history, it has now become a transcendental condition *of* the possibility of the "projected unity" (A647/B675) of nature, of the world as described by scientific theory, or by a nest of such theories, forming a Lakatosian "research-programme": Scientific reason no longer "begs but commands" (A653/B681).

The significance of 'reduction' will now be clear. The philosophical grammar, or 'basic phenomenology' of things (or of 'nature') can only be determined with 'certainty' if their ontology has first been generated *via* a constitutional process of 'realization', in the sense described. Only in this way can we be *certain* that the "object conforms to the concept" (*cf.* Bxvii).

We had asked: how can we define (and thus save) the notion of a 'rational justification' of our methodological scheme? And our answer is: Only if we treat any object ("nature", "the order of nature", *etc.*) as though its "real possibility" was conditional upon the provision of a given set of enabling conditions; conditions which – in cases of success – are thereby "validated" likewise. But such enabling conditions can be clamped on 'a priori' (thus providing a 'realization') only if the object has been previously 'reduced'. The prior 'reduction of the world' ensures that no questions can be asked concerning this world, nor any hypothetical answers be assumed, or even postulated, concerning its constitution, outside a process of 'realization'; apart from it, one cannot ask whether the world obeys a principle of uniformity, or of simplicity; or whether – a central problem for Kant's time – a 'natural classification' applies to it, into genera and species. Indeed this last case affords us one of the most instructive insights into the whole Kantian approach, containing the very essence of his philosophy of scientific theory construction.

If we regard the system of nature, so argues Kant, as having the status of a "reality in itself" (A679/B707), "its unity is really surrendered as being quite foreign and accidental to the nature of things, and as not capable of being known from its own universal laws" [i.e. its methodological structure] (A693/B721). We could then never know whether, e.g. a classification into genera and species was a real possibility, since ultimately all may be chaotic "variety". Hence, in order to be able to "apply" the methodological "principle of genera" to nature, we must 'presuppose a [corresponding] transcendental principle, in accordance with which *homogeneity* is necessarily presupposed in the manifold of possible experience; for in the absence of homogeneity, no empirical concepts [e.g. of genera] and therefore no [systematic] experience would be possible' (A654/B682). We thus have the position that the required "homogeneity" (and consequently classifiability) of nature can only be "presupposed", i.e. taken as defining nature in the required manner, if apart from such a presuppositional framework one could *neither affirm nor deny* anything concerning its structure. (For it would then be reduced to a mere O_t). It is in this sense that the "order" of nature must first have been subjected to a 'reduction', so that only *then* can a process of 'realization' (here: application of the principle of genera) render a natural classification "possible". In this sense one may say, with Kant, that proofs of possibility are "apodeictically certain in every hypothesis" concerning empirical nature (Kant 1974: 93; cf. 1970: 4).

Does this avoid the sceptical position of some contemporary

Kant, Immanuel. 1968. Enquiry concerning the clarity of the principles of natural theology and ethics. In *Kant Selected Pre-Critical Writings*, ed. G. B. Kerferd and D. E. Walford, pp. 5–35. Manchester.

Kant, Immanuel. 1970. *Metaphysical Foundations of Natural Science*, trsl. James Ellington. Indianapolis and New York.

Kant, Immanuel. 1974. *Logic*, trsl. R. Hartman and W. Schwarz. Indianapolis.

Kuhn, T. S. 1962. *The Structure of Scientific Revolutions*. Chicago.

Lakatos, Imre. 1970. Falsification and the Methodology of Scientific Research Programmes. In *Criticism and the Growth of Knowledge*, ed. I. Lakatos and A. Musgrave, pp. 91–196. Cambridge.

Laudan, Larry. 1977. *Progress and its Problems*. London.

Meyerson, Émile. 1930. *Identity and Reality*. London.

Putnam, Hilary. 1978. *Meaning and the Moral Sciences*. London.

Quine, W. V. 1969. *Ontological Relativity and other Essays*. New York.

Rorty, Richard. 1979. Transcendental Arguments. Self-Reference, and Pragmatism, in *Transcendental Arguments and Science*, ed. P. Bieri, R.-P. Horstmann and L. Krüger, pp. 77–103. Dordrecht.

Schleiden, M. J. 1849–50. *The Principles of Scientific Botany, together with a Methodological Introduction as a Primer for the Study of the Plant* 2 vols., 3rd ed. Leipzig.

Sellars, W. 1963. *Science Perception and Reality*. London.

Stallo, J. B. 1885. *The Concepts and Theories of Modern Physics*, 2nd ed. London.

Whewell, William. 1967. *The Philosophy of the Inductive Sciences*, ed. G. Buchdahl and L. L. Laudan, vols. V and VI (Parts I and II). London.

2   *The possibility of absolutism*

NICK JARDINE

1 It is a familiar thought that through progress in the natural sciences we have achieved ever more impersonal and objective representations of the world, representations which portray it ever more nearly as it is, not as it appears from some particular vantage point. And it is a familiar hope that there is no intrinsic limit to this growth of objective knowledge. These I shall call *absolutist* attitudes.[1] Equally familiar is the sobering suspicion that absolutism is unwisely presumptuous about our cognitive capacities. Reflection on the limitations of human nature, perhaps reinforced by consideration of the extent to which the capacities of the human senses and intellect are conditioned by our biological particularity, may convince us that there are certain vantage points to which we are forever tied by our humanity and hence cannot hope to transcend in our scientific theories. and reflection on the diversity of human nature, on the variety of actual and possible human interests, methods of inquiry and images of man and his place in the world, perhaps reinforced by consideration of the extent to which such matters are conditioned by and reflect the disparate forms of human society, may convince us that our scientific theories cannot transcend even the special vantage points of their creators and protagonists, let alone those of the human race. These are *relativist* attitudes. The relativism which sees each possible human theory as limited by the human condition itself may be called *external* relativism; that which sees each possible human theory as limited by the special

---

[1] Following Williams (1978: 64–6). 'Realism' has too many other connotations, and 'objectivism' is misleading, suggesting as it does a thesis about the existence of public criteria for appraisal of theories.

vantage points of its creators and protagonists may be called *internal* relativism.

Clear though the conflict between absolutist and relativist attitudes to the natural sciences appears, it is notoriously difficult to consolidate the absolutist attitudes into intelligible theses. For it is hard to counter the charge that in appealing to the notion of representation of the world as it is the absolutist appeals to a notion we do not possess and cannot acquire.

2   We may reformulate the absolutist attitudes as theses about the truth of scientific theories, truth being identified with representation of the world as it is, and transcendence of vantage points being identified with avoidance of sources of systematic error and partiality. This is, I think, a first step in the direction of intelligibility. What should be the next step?

If the way in which a true theory represents the world could be specified, if an account of truth as correspondence could be spelled out, then the task of rendering absolutism intelligible would surely be greatly facilitated. The hope of establishing such an account of truth has motivated much of the recent spate of work on the theory of reference, and the prospect of wielding such an account as a weapon against the relativist lies behind much of the recent obsession with reference displayed by philosophers of science. Suppose we had an explicit theory of reference, one which without appeal to other semantic notions determined the referential scheme of the language of population P, for arbitrary P. And suppose we could construct Tarskian semantic apparatus of sufficient power to deal with natural languages. Then where Tarski was able to define truth in a language for a restricted class of artificial languages, we would be able, so the story goes, to define truth in the language of population P, for arbitrary P. We would then have a correspondence theory of truth.[2]

Many specific objections to this programme have been raised. For example, Rorty (1976) argues convincingly that philosophical discussions of reference conflate a number of distinct questions we customarily raise in interpreting the utterances of others – hence the apparent conflict of intuitions that gives rise to conflicting theories of reference. Further, each of the main theories of reference faces

---

[2] For a clear presentation of this programme see Field (1972).

severe internal difficulties. Causal theories are hard put to specify the kinds of causal chains that preserve reference without begging the question. Fregean theories are hard put to say how strictly *F*s must satisfy speakers' beliefs about *F*s; if they require only a loose fit they are threatened with prevalent indeterminacy of reference; if they demand a tighter fit they are faced with prevalent empty reference. And consensus theories are hard put to say who is competent to have a say on questions of classification without appealing to the knowledge a speaker has, and hence, question-beggingly, to the notion of truth. Finally, there is the suspicion that any theory of reference that appeals to the contents of speakers' beliefs fails to provide the proper basis for a correspondence theory of truth since it appeals to entities that can be sufficiently precisely individuated only by semantic criteria.

But there is a much more general objection. Underlying the suspicion that the notion of representation of the world as it is may be unintelligible is the view that we can have no conception of the world that transcends all possible representations of the world; that though maybe the world can, *pace* Ayer (1974: 235), "be prised away from our manner of conceiving it", it cannot be prised away from every possible manner of conceiving it. To attempt to specify absolutely a relation of correspondence between the world and representations of the world is futile. This view is, I think, inescapable. If absolutism is to be rendered intelligible we must, at the outset, be content with the notion of truth (and its cognates, reference, consistency, equivalence, *etc.*) from the standpoint of a representation of the world, that is, a theory.

The thought naturally arises that absolutism may be rendered intelligible by specifying some ultimate theory, U, to serve as the standard of truth. The futile task of saying in what correspondence with the world consists may then be set aside, for the claim that a theory represents or partially represents the world may be understood as the claim that from the standpoint of U it is true or partially so. What could constitute such an ultimate theory? The traditional answer was God's knowledge of the world. A humanist substitute for God's knowledge is the limit of human inquiry, the theory that would be attained were human inquiry to continue forever.[3] This

---

[3] Sellars is a notable exponent of this kind of absolutism. It is generally attributed to Peirce; but if Thompson's (1978) interpretation is right Peirce espoused a form of absolutism closer to the one to be presented here.

attempt at intelligibility is, I shall argue, a failure. But it is, as we shall see, an instructive failure.

Is the notion of a limit of inquiry itself intelligible? Consider an infinite series of theories, $\Theta = T_1 \ldots T_i \ldots$, taking 'theory' broadly to include conjunctions of theories in the customary usage. To find a sense in which the series has a limit we must find a sense in which it converges. In so doing we cannot, on pain of circularity, appeal to a measure of degree of approximation of a theory to the truth; we must make do with a measure of degree of approximation of one theory to another. Convergence in the series may then be defined as follows. Take any degree of approximation however close, ε, say; by choosing a large enough value of i we can ensure that T_i approximates to each $T_{i'>i}$ to within ε. Now consider an infinite series of collections of theories, $I(K) = C_1 \ldots C_i \ldots$, the ith member of the series consisting of all theories entertained by inquirers of kind K at time i. We call $I(K)$ an *inquiry series*. A series of theories Θ is contained in an inquiry series $I(K)$ if, for all i, T_i belongs to C_i. Θ is a maximal convergent series of theories contained in $I(K)$ if it is convergent and there is no convergent series of theories Θ' contained in $I(K)$ such that, for some i, T_i is a conjunct of T_i' ($\neq T_i$). $I(K)$ is convergent if it contains just one maximal convergent series of theories. Its limit is the limit of that series of theories.

As a partial clarification of the 'limit of inquiry' version of absolutism we offer the postulate that the series of collections of theories that have been entertained and would be entertained were human inquiry to continue indefinitely, $I(H)$, is a convergent series in the sense just outlined, its limit being a representation of the world as it is.

There are at least three salient objections. (1) It is doubtful whether generally defined numerical measures of degree of approximation of one theory to another are to be had. It is surely senseless to assert, for example, that Newtonian mechanics is twice as close to special relativistic mechanics as Mendel's factor theory is to Morgan's mathematical genetics. (2) To the extent that we can make sense of the notion of degree of approximation of one theory to another, the history of science does not suggest that human inquiry has generated the initial segment of a convergent series of theories. We can, indeed, detect a successive approximation of past theories to our own. But the rate of approximation seems to have been uneven, recent "conceptual revolutions" having been no less drastic than earlier ones. Yet to make a *prima facie* case for existence of a limit of human inquiry we would have to detect diminishing returns with the passage of time, a general deceleration of progress as estimated from the standpoint of our theories. Perhaps the convergence some authors have perceived in the history of science is an illusion generated by hasty extrapolation from cases in which estimates of some parameter (*e.g.* the distance of the Sun) according to successive past theories converge on our estimate. Or maybe the illusion arises through conflation of two distinct phenomena, the successive approximation of past theories to our own and the progressive unification of domains of inquiry through microreduction. (3) Identification of representation of the world as it is with such a limit of inquiry fails to capture the absolutist position. For under this interpretation absolutism would be compatible with relativism. Nothing in the limit of inquiry version prevents the external relativist from identifying the postulated limit of inquiry as a representation subject to sources of systematic error and partiality that we cannot by virtue of our humanity transcend. And the internal relativist is free to maintain that the limit of inquiry is conditioned by the particular vantage points of the creators and protagonists of particular theories – Would not the series have tended to a different limit had systematic theorizing about the nature of the world originated in China rather than Ionia, or had the founders of empirical psychology not been in the grip of foundationalist epistemologies?

The first objection can perhaps be evaded; for analogues of convergence can be constructed without assuming either that the measure of approximation of one theory to another is numerical or that it is defined for all pairs of theories.[4] The second is more serious. It is doubtful whether under any plausible measure of degree of approximation the history of science would yield the required evidence. The third is, I think, fatal. What can be said about the nature of the limit of human inquiry that would block relativist interpretations of the kinds just sketched? Merely to credit the limiting theory with superlative degrees of all that we most cherish in a scientific theory, with the uttermost universality, coherence, clarity, formal and ontological parsimony and sustained predictive

---

[4] This would require considerable ingenuity, for it seems that we are entitled at best to assertions of the form 'T_i is nearer T_j than is T_k'.

power, does nothing to block these relativist readings. Rather what is needed is a limiting theory that explicitly denies them, that would itself catalogue all possible sources of systematic error and partiality and would enable us to prove that it itself transcends them by proving that the methods by which it is warranted are immune to them all. Where the history of science, taken at face value, fails to support the claim that there exists a limit of human inquiry, it positively militates against the existence of a limit of this sort. From the standpoint of our theories we can, to be sure, see how sources of systematic error and partiality, at first unrecognized, have been detected and then eliminated or controlled. But far from providing inductive grounds for supposing that the process of transcendence of sources of error would, were inquiry to continue indefinitely, one day be demonstrably complete, it suggests that there is no end to this process of recognition and subsequent avoidance or evasion of sources of error.

The limit of inquiry version of absolutism can be seen as an attempt to make absolutism measure up to verificationist standards of intelligibility. The absolutist claims that some at least of our theories and their components represent the world as it is (or approximately so). Well-known arguments for the revisability in principle of each component of our theories rule out conformity to the strict verificationist standard, that which would demand specification of an actual procedure by which we can discern among our scientific beliefs just those which thus represent the world. But the limit story does promise to meet a more relaxed verificationist standard, that which allows sense to be conferred by specification of a hypothetical verification procedure, one which, in Dummett's (1973: 465) words, involves 'perceptual or mental operations which transcend our capacities . . . derived by analogy with those we do perform'. By analogy with an operation we do perform, construction of a theory and assessment of the truth values of other theories and their components from its standpoint, the absolutist derives a hypothetical verification procedure to confer sense on his claim, construction of an ultimate theory through endless inquiry and assessment of the truth values of its predecessors and their components from that ultimate standpoint.

That absolutism cannot measure up to such verificationist standards is hardly surprising. For there is tension between the absolutist postulation of a world as it is, a world independent of the perceptions and cogitations of inquirers, and the postulation of an actual or hypothetical procedure whereby it could be known of some particular representation that it represents the world as it is. The independence of the world is not threatened by the claim that our methods of inquiry are reliable enough to enable us in some measure to represent it as it is, to have knowledge of it. But it is surely threatened by the claim that we can, or could hypothetically, know of some particular representation that it constitutes knowledge. Like the attempt to ground absolutism in a correspondence theory of truth this attempt demands too much accessibility of the world, more than the history of science entitles us to expect and more than the absolutist can afford to countenance.

3 Can a better balance between the objectivity and the accessibility of the world be struck? Can we prise the world away from our actual manner of conceiving it without tacking it down again to a particular hypothetical manner of conceiving it?

Taken at face value the history of science shows successive approximation of past theories to our own, but not the diminishing returns that the limit of inquiry version of absolutism requires. Further, from the standpoint of our theories we can see how sources of error and partiality, at first unrecognized, have been detected and then transcended, though we cannot discern the tendency towards exhaustive recognition and transcendence of sources of error that the limit of inquiry version demands. Let us try to formulate absolutism as a simple extrapolation from this historical evidence.

As before, we consider an infinite temporally indexed series of theories, $\Theta = T_1 \ldots T_i \ldots$, again using 'theory' in the broad sense to admit conjunctions of theories in the customary sense. We want to define three properties of such a series: *coherence*, the theories at earlier stages being true (or approximately so) from the standpoint of those at later stages; *cumulativeness*, the theories at earlier stages being contained in those at later stages as components or special cases; and *transcendence*, sources of error and partiality recognized at earlier stages being avoided or controlled by the theories which appear at later stages. As before we are entitled only to use the notion of truth (and its cognates, consistency, equivalence, *etc.*) from the standpoint of a theory.

We say that a theory or component of a theory, T, is true (or approximately so) from the standpoint of Θ if there is an i such that for all $T_{i'\geqslant i}$ in Θ, T is true (or approximately so) from the standpoint of $T_{i'}$. Similarly for falsity from the standpoint of Θ (or which approximate truth from the standpoint of Θ is a special case). A theory or component of a theory that is neither true nor false from the standpoint of Θ has indeterminate truth value from the standpoint of Θ. Θ is *coherent* if, for all i, T_i is true (or approximately so) from the standpoint of Θ. A theory or component of a theory T is new at t from the standpoint of T' if from the standpoint of T' it is not equivalent (or approximately so) to any theory or component of a theory proposed before t. It is new at t from the standpoint of Θ if there is an i such that for all $T_{i'\geqslant i}$ it is new at t from the standpoint of $T_{i'}$. Θ is *cumulative* just in case there is no i such that there does not exist $T_{i'>i}$ containing theories or components of theories that are from the standpoint of Θ both new at i' and true (or approximately so) – that is, there is no end to the introduction of material that is both new and at least approximately true from the standpoint of the series.

An appropriate definition of transcendence is not so easily come by. Again we cannot appeal to absolute notions of error and partiality and their avoidance: all must be relativized to a series of theories. But clearly it is not enough to require, by analogy with the definition of cumulativeness, that there be no end to the process of avoidance of what are, from the standpoint of a series of theories, sources of systematic error and partiality. For that is consistent with there being sources of systematic error and partiality, recognizable from the standpoint of the series, that cannot be overcome at any stage in the series, a fatal concession to the relativist. On the other hand, to require there to be a stage in the series at which all sources of error and partiality recognizable from the standpoint of the series are demonstrably overcome is, as we have seen, to require too much. So we take a middle line. Let E be a source of error or partiality from the standpoint of Θ if there is an i such that, for all $T_{i'\geqslant i}$ in Θ, E is a source of systematic error or partiality from the standpoint of $T_{i'}$. Likewise, E is transcended by T from the standpoint of Θ if there is an i such that for all $T_{i'\geqslant i}$ in Θ, T avoids E from the standpoint of $T_{i'}$. Θ is *transcendent* if there does not exist E such that there does not exist T_i in Θ that transcends E from the standpoint of Θ. This may be seen as a powerful coherence condition. A transcendent series of theories is one from the standpoint of which there are no ineluctable sources of systematic error or partiality.

Now consider, as before, an inquiry series; that is, an infinite series of collections of theories, $I(K) = C_1 \ldots C_i \ldots$, the ith collection consisting of all theories entertained by inquirers of kind K at time i. We say that $I(K)$ is coherent if it contains just one maximal coherent series of theories, Θ. A theory that is true (or approximately so) from the standpoint of Θ is true (or approximately so) from the standpoint of $I(K)$. We say that $I(K)$ is cumulative if Θ is cumulative; transcendent if Θ is transcendent.

We are now in a position to offer our version of absolutism.

A **i.** A theory or component of a theory represents the world (approximately) as it is, if it is (approximately) true from the standpoint of a coherent, cumulative and transcendent inquiry series, $I(K)$.

    ii. The human inquiry series, $I(H)$, is coherent, cumulative and transcendent.

Perhaps **A** should be called 'human absolutism'. Absolutism about Martian science, the natural stance for Martian absolutists, is obtained by substituting the Martian inquiry series for the human inquiry series in **A ii**.

A attempts to strike the right balance between the objectivity and the accessibility of the world, to articulate a thoroughgoingly fallibilist absolutism. By liberating the world from any particular representation it purports to secure the objectivity that absolutism demands. And by tying the world to the fortunes of possible types of inquiry it purports to achieve that modicum of accessibility that intelligibility demands.

If **A** is to put absolutism on the cards it must at least be intelligible, inconsistent with the various relativisms, immune from *a priori* demolition, and tolerably well supported by evidence germane to the issue. No systematic defence will be attempted here. In the following sections I shall merely sketch what seem to me to be the most pressing objections on these scores and adumbrate some defensive strategies.

4 We have taken as primitive the notion of truth from the standpoint of a theory. Is this notion itself well-defined? Suppose that under whatever are the constraints on interpretation of one theory from the standpoint of another there is no unique best interpretation, either because several divergent interpretations satisfy the constraints perfectly, or because no interpretation satisfies them perfectly but there are divergent optimal imperfect solutions. Then surely we must concede that the truth value of one theory or component of a theory from the standpoint of another is ill-defined. Our primitive notion should be truth of one theory or component of a theory from the standpoint of another under a given interpretation. Let us make the required concession. To reinstate our original primitive notion we say that a theory or component of a theory is true from the standpoint of another theory just in case it is true from the standpoint of that theory under all admissible interpretations.

A receives *prima facie* support if from the standpoint of our theories the history of science reveals both accumulation of knowledge and successive transcendence of sources of systematic error and partiality. Let us call this 'the required evidence'. The concession just made begs off a serious objection to the intelligibility of **A**, but at the cost of rendering the reading of the history of science that lends credence to **A** open to serious doubt. Thus each of the following would if substantiated deprive **A** of its support. (1) Under no admissible interpretation does the history of science yield the required evidence. (2) There are some admissible interpretations under which the history of science does not yield the required evidence. (3) Under each admissible interpretation the history of science yields the required evidence, but what is common to all the various interpretations does not itself constitute an interpretation that yields the required evidence.

(1) has little initial plausibility. It is, however, an immediate consequence of a certain kind of theory of reference. For if we insist, as does one version of the Fregean theory, that the referents of theoretical terms must satisfy all the high-level postulates in which the terms figure, then from the standpoint of our theories the theoretical terms of past theories will almost always have null reference. From the standpoint of our theories most past theories will be false through-and-through because about nothing. Our earlier strictures against the very idea of a theory of reference may be used to deflect this line of attack.

(2) is slightly more plausible. It may well appeal to those historiographers who suppose that although the kind of reading of the history of science which yields support for absolutism is all very well for certain purposes – explaining the growth of technology or inspiring budding scientists, for example – other less Whiggish

kinds of reading are better for other purposes, relating past scientific theories to their contemporary intellectual and cultural *milieux*, for example. There is an obvious line of defence against this. As Grandy (1973) has argued, whilst we should reject the naive "principle of charity", a sophisticated principle which demands that we minimize the attribution of inexplicable error is a central constraint on interpretation. And this suggests that whilst we cannot hope to justify *a priori* the claim that a proper interpretation of the history of science reveals accumulation of knowledge and progressive transcendence of sources of systematic error, we can justify *a priori* the claim that *if* any such interpretation is admissible *then* interpretations which do not have this character are inadmissible.

(3) is the really threatening thesis. If true it would explain why the history of science appears to yield the required evidence. And unlike (1) and (2) it is supported by putative examples. Thus Field (1973) has argued that there are at least two equally good ways of exhibiting Newtonian mechanics as an approximation to special relativistic mechanics,[5] and Sklar (1976) has presented a similar argument for the case of classical thermodynamics and statistical mechanics. Whether these are cases of (3) is a moot point; I think not. Each of the interpretations is, indeed, consistent with the required type of interpretation of the history of science, for under each the move from the reduced to the reducing theory can be seen as involving both growth of knowledge and transcendence of sources of systematic error. Further, from the standpoint of the reducing theory the different interpretations assign different referents to some of the terms of the reduced theory and different truth values to some of its components. But the overlap between the interpretations is substantial. They make almost exactly the same components of the reduced theories come out approximately true and provide almost exactly the same explanations for the predictions of the reduced theory having been, within the limits of accuracy of measurement available to its protagonists, largely successful. So it seems that, *contra* (3), the intersection of these interpretations is itself an interpretation under which the move from the reduced theory to the reducing theory can be seen to involve both growth of knowledge and transcendence of sources of systematic error. I suspect that a general theory of interpretation would allow *a priori* dismissal of (3) along the line sketched for (2).

[5] Earman (1977) and Yoshida (1977: Appendix) have denied Field's claim.

At best we have shown that indeterminacy of interpretation does not obviously entail either the unintelligibility of **A** or the invalidity of the historical evidence that lends support to **A**. In so doing we have issued a large promissory note drawn on the theory of interpretation, a promissory note that cannot be redeemed here.

5   Where indeterminacy of interpretation threatens to deprive **A** of its historical support, another Quinean thesis, underdetermination of theory by all possible evidence, appears to strike more directly at **A**. However, care is needed in formulating the underdetermination thesis if the conflict is to be properly located. The thesis is generally expressed as follows. There exist n-tuples of theories that are mutually inconsistent and are such that there exists no evidence that would resolve the issue. As it stands, this looks like a challenge to the adequacy of **A** as a version of absolutism. For it is tempting to reason as follows: Suppose the human inquiry series, $I(H)$, is coherent, cumulative and transcendent, but that some underdetermined theory, T, is from its standpoint true. Then, by hypothesis, T, represents the world as it is. But T might just as well be replaced by one of its empirical equivalents, T', in which case we would have a different human inquiry series, $I(H)'$, from whose standpoint T' would be true and T false. So truth from the standpoint of $I(H)$ cannot be representation of the world as it is, only representation of the world as it is from the standpoint of an arbitrary choice of theory.

This challenge is illusory. The protagonist of the radical underdetermination thesis is no more entitled to employ absolute notions of inconsistency, evidence and resolution of theoretical conflict than is the absolutist. So let us reformulate the thesis in a suitably relativized form.

U.   There exist n-tuples of theories that are from the standpoint of $I(H)$ inconsistent and such that from the standpoint $I(H)$ there exists no evidence that would resolve the issue.

(Here inconsistency, evidence and resolution of theoretical conflict from the standpoint of $I(H)$ are defined according to the schema used in section **3** to define truth from the standpoint of an inquiry series.) Perhaps we should call this 'the human underdetermination thesis'. Martian absolutists will presumably be perturbed by a version in which $I(\text{Martian})$ is substituted for $I(H)$.

There is a crucial ambiguity in **U**. On the one hand, we may allow that a theoretical conflict can be resolved by a body of evidence that whilst entailed by each of the conflicting theories supports one of them better than the others. By this standard one may, for example, claim that the observations of apparent stellar and planetary coordinates available in the sixteenth century resolved the issue between the Copernican and Ptolemaic systems, on the grounds that whilst for suitable choice of parameters both systems predict those observations equally well, the Copernican system having fewer independent parameters is better supported by them (*cf.* Rosenkrantz, 1977: Ch. 7). On the other hand, we may require that for a body of evidence to resolve a theoretical conflict it must be entailed by but one of the conflicting theories. In this sense the issue between the Copernican and Ptolemaic systems was not resolved by kinematic evidence until the observation of stellar parallax in the nineteenth century. Let us call the two versions of **U**, that which employs the stringent notion of resolution of conflict of evidence and that which employs the more relaxed notion, the *weak* and *strong* underdetermination theses, **WU** and **SU**, respectively.

SU is inconsistent with **A**. Acceptance of a theory on inadequate grounds is a source of systemic error, as is rejection of a theory on inadequate grounds. Suppose $I(H)$ contains a strongly underdetermined theory T. If T is true from the standpoint of $I(H)$, then from the standpoint of $I(H)$ the first source of error is never transcended. If T is false from the standpoint of $I(H)$ then from the standpoint of $I(H)$ the second source of error is never transcended. There is, of course, a third possibility: T may have indeterminate truth value from the standpoint of $I(H)$. And that, it may be suggested, is how it should turn out for the inquiry series of rational beings.[6] But again $I(H)$ cannot be transcendent, for failure to resolve real issues is a source of partiality, and since T and its empirical equivalents are from the standpoint of $I(H)$ inconsistent the issue between them is from the standpoint of $I(H)$ a real one. So if **SU** is true, $I(H)$ is not transcendent; hence **A** is false.

WU does not conflict directly with **A**. And that is fortunate, for the case for **WU** can be made very strong indeed; see, for example, Van Fraassen (1976). It may, however, be argued that **WU** weakens **A** indirectly by forcing its protagonists to assume that as well as

<hr />

[6] *Cf.* Quine's (1975: 328) remarks on the proper reaction to an encounter with empirically equivalent theories.

bare predictive success such criteria as structural simplicity, ontological parsimony and intelligibility are to be taken into account in assessing the evidential support of a theory. And this, it may be urged, reinforces the suspicion that truth from the standpoint of $I(H)$ is to be equated not with representation of the world as it is, but with representation of the world as it is from the standpoint of a particular methodology. The confines of this paper do not permit any general defence against this powerful line of attack. But there is an obvious *ad hominem* rebuttal. The protagonist of **WU** concedes the validity of predictive success as a criterion for resolution of theoretical conflict. But if that is conceded the way is open for vindication of other canons of evidential support by appeal to their success as predictors of predictive success in theories.[7] Thus, for example, the use of paucity of independently adjustable parameters as a criterion is vindicated by the eventual confirmation of the Copernican prediction of stellar parallax.

What of the truth of **SU**? Non-standard theories of space, time and space–time of the kinds intimated by Poincaré and constructed by Reichenbach and Grünbaum are perhaps the most promising candidates. Yet Glymour (1977) and Friedman (1980) have argued convincingly that these, whilst they may well instantiate **WU**, do not instantiate **SU**. In the absence of good putative examples there is, I submit, no case to answer.

6 The objections to **A** just mooted would, if successful, establish forms of internal relativism. What of the lines of attack on **A** conducive to external relativism? Of a plethora of candidates I shall consider just two which are, I think, peculiarly ominous.

First there is the claim, familiar to readers of Bergson and Husserl, that our natural sciences have an essential incompleteness, one for which they hold out no promise of rectification. As Einstein once remarked, 'you cannot expect scientists to give you the taste of soup'. It is, so the argument goes, undeniable that there are facts about the quality of the sensory experiences had by each of us, and presumably the same is true of bats and the sentient inhabitants of other galaxies. Such facts do not figure in our scientific theories nor is there reason to suppose that the successors to our theories will come to terms with them.

---

[7] A line of thought developed in detail by Rescher (1977).

It is tempting to answer as follows. These "facts" that the natural sciences leave out are not aspects of the way the world is, but rather, like the values things have for each of us, ways the world appears from particular vantage points. Since in our natural sciences we are concerned only with the way the world is, our natural sciences cannot be expected to include them. The temptation is to be resisted, for it misses the point of the objection. What is at issue is the postulate of unlimited transcendence of vantage points by the human inquiry series. Transcendence of a vantage point is recognized when we have a theory from whose standpoint it is possible to explain why the world appears as it does to those who occupy that vantage point. The spirals in which the planets appear to the earthbound observer to move are not components of the world and accordingly do not figure in our astronomical theories, but the fact that they thus appear to him surely is, and it is incumbent on our astronomical theories to explain it, as indeed they do. Likewise, though the taste mushroom soup has for us is not a component of the world and hence cannot properly figure in any scientific theory, if the human inquiry series is transcendent we should one day be able to explain why mushroom soup tastes to us as it does.[8] The real force of the objection is now clear. To achieve that kind of transcendence, the objector will insist, we would need, in Nagel's (1979: 213) words, 'a general conception of experience which admitted our own subjective viewpoint as a special case'. As Nagel goes on to observe, 'this is completely beyond us and will probably remain so as long as human beings continue to exist'.

In attempting to counter this line of attack the absolutist is surely entitled to deny that he is committed to the possibility of a general theory of experience, and hence to refuse to be drawn into futurological speculation about the form such a theory might take.[9] The postulate of transcendence of the human inquiry series does not, after all, require that a theory which explains all aspects of the experience of all possible sentient beings be on the cards, only that there be no such aspect that will forever resist explanation. But if absolutism is not itself to be an ungrounded futurological speculation he must, I think, deny that all such explanations are 'completely beyond us'. He must deny that our science offers us no prospect

[8] *Cf.* Williams (1978: 295).
[9] As, for example, Sellars (1963: Ch. 1) is drawn when he makes definite predictions about the content of any theory adequate to explain the qualitative aspects of our sensory experience.

whatever of explanations of the content of the experience of sentient beings. Of course, we are nowhere near to having complete explanations of any aspect of our experience. But combining physics with physiology we are able to provide partial explanations of such aspects of our experience as our sensations of hue, brightness, distance, orientation and movement. Certainly one is suspect who belittles such explanations out of hand as dealing only with peripheral aspects of experience, for the onus is on him to specify the aspects of our experience that will forever resist explanation. To show that such partial explanations constitute evidence against the claim that there are aspects of our experience which the natural sciences cannot hope ever to explain, as the product of our constitution and situation in the world, is an undertaking beset with well-known difficulties. For all that, it is, I think, one that the absolutist is bound to attempt.

Having conceded the gravity of the "taste of soup" objection to absolutism, let us turn to the possibility of alien intelligences. They too may conduct scientific inquiries. So we must take into account not just the series of collections of theories that would be entertained were human inquiry to continue indefinitely, $I(H)$, but also the inquiry series that would arise from indefinitely prolonged scientific inquiry by other life-forms, $I(Martian)$, $I(Venerian)$, and so on.

At first sight this possibility appears to cast doubt on the adequacy of **A** as a formulation of absolutism. For one may speculate as follows. Suppose both of two inquiry series, $I(M)$ and $I(V)$, say, are coherent, cumulative and transcendent. Then, according to **Ai**, representation of the world as it is is defined for each of them. But what conceivable grounds can there be for supposing that these definitions are congruent, that whatever is true from the standpoint of $I(M)$ is true from the standpoint of $I(V)$, and *vice versa*? Yet, in the absence of such grounds, truth from the standpoint of a coherent, cumulative and transcendent inquiry series, $I(K)$, is surely to be identified not with representation of the world as it is, but with representation of the world as it is for beings of kind K.

This challenge is illusory. As in the case of radical underdetermination of theory, the real challenge is to the truth not the adequacy of **A**. Take any two inquiry series, the Martian and the Venerian say, and consider a hypothetical encounter. We first adopt the Martian standpoint. There are three possibilities. (M1) Martians are eventually able to understand Venerian science. It turns out to be surpassed by, to be equivalent to, or to surpass their own; or it is a mixture of components of these types. (M2) Martians are eventually able to understand Venerian science. It turns out to include some theories that are inconsistent with, but strongly empirically equivalent to, their own. (M3) Come what may Venerian science remains inaccessible to Martians. If we adopt the Venerian standpoint exactly similar possibilities (V1–3) arise. None of these possibilities impugns the adequacy of **A** as a version of absolutism. If (M1) and (V1) hold, then there are good grounds for supposing that whatever is true from the standpoint of the Martian inquiry series is true from the standpoint of the Venerian inquiry series, and *vice versa*; and hence that if the two series are coherent, cumulative and transcendent they yield congruent definitions of representation of the world as it is. If (M2) or (V2) holds, there is again no challenge. For the thesis of radical underdetermination of theory now holds from the standpoint of one of the two series, so one of the two series is not transcendent, and representation of the world as it is is not defined with respect to it. If (M3) or (V3) holds there is yet again no challenge to the adequacy of **A**. For inability of a kind of inquirers to understand the fruits of the inquiries of beings of another kind, beings whose cognitive capacities may vastly outdistance their own, is a source of partiality in their theories. So again one of the two inquiry series is not transcendent, and representation of the world as it is is undefined with respect to it. If, as I maintain, the various combinations of (M1–3) with (V1–3) exhaust the relevant possibilities, the challenge to the adequacy of **A** as a version of absolutism is dispelled.

The possibility of alien intelligences can be seen, however, to raise two kinds of challenge to the truth of **A**. One is a new version of a familiar challenge. Should some alien life-form turn out to hold theories that are strongly underdetermined with respect to our own, then, as shown in section **5**, $I(H)$ is not transcendent, so **A** is false. The other is altogether new. Should the science of some alien life-form, Martians say, turn out to be inaccessible to us, $I(H)$ is not transcendent, so **A** is false. This new challenge is a powerful one. Whatever confidence we have in the translatability of other human languages into our own is grounded not merely in the belief that there are *a priori* constraints on the patterns of belief, desire and action that can be manifested by rational agents, including Martian rational agents, but also in the conviction that the experiences had by others have much in common with our own. The latter belief is presumably unwarranted if the others are Martians. So much the worse for the prospects of access to Martian science.

Worse still, the "taste of the soup" argument and the "Martian inscrutability" argument are mutually reinforcing. To meet the taste of the soup argument we must deny that there are aspects of Martian experience that we can never hope to explain. But to explain any aspect of Martian experience we must surely have some initial conception of the nature of Martian experience. This we cannot hope to acquire by *Verstehen*, by empathetic extrapolation from the nature of our own experience. But we might hope to acquire it if only we could understand what Martians say about their experience. This hope is doomed if the Martian inscrutability argument succeeds. Conversely, to meet the Martian inscrutability argument we need, again, some initial conception of the nature of Martian experience. We might hope to acquire such knowledge by deploying theories which, on the basis of information about the constitution and situation in the world of Martians, would deliver predictions about the nature of their experience. This hope is doomed if the taste of the soup argument succeeds.

7   Our aim was to find a version of scientific absolutism that is both intelligible and has a sporting chance, being supported by a reasonable interpretation of the history of science and not being readily defeasible *a priori*. **A** does, I think, capture the central tenets of a fallibilist absolutism, an absolutism which claims objective progress for the natural sciences whilst abandoning all hope of attainment of completeness or certainty. Its status as a version of absolutism is confirmed by its vulnerability to a variety of arguments conducive to relativism. That it is not too readily defeated I have tried to show by sketching lines of defence against some of those arguments. The intelligibility of **A** remains, I think, in the balance. The arguments for external relativism have strained to the limit the notion of transcendence, of diagnosis from the standpoint of one theory of sources of systematic error and partiality to which other theories are subject. And our uncritical use of the notions of theoryhood and of approximate truth of one theory from the standpoint of another raises further doubts on the score of intelligibility.

Absolutism can surely be rendered with greater precision. In this paper I have uttered large promissory notes drawn on the theory of interpretation, notably in contexts where others have uttered promissory notes drawn on the theory of reference. From a general theory of interpretation there would, I believe, ensue much clarification of the dubious notions used in the present articulation of absolutism. But I suspect that even then absolutism would remain at the frontier of intelligibility. Perhaps only the very progress of science that absolutism promises us should we survive and keep on trying can render absolutism fully intelligible, by gradually revealing to us how it is possible for us to have the knowledge which we reveal in our theories and in our interpretations of the theories of others, the knowledge which the absolutist arrogates to us.[10]

University of Cambridge

REFERENCES

Ayer, A. J. 1974. *The Central Questions of Philosophy*. London.
Dummett, M. A. E. 1973. *Frege: Philosophy of Language*. London.
Earman, J. 1977. Against indeterminacy, *The Journal of Philosophy* **74**, 535–8.
Field, H. 1972. Tarski's theory of truth, *The Journal of Philosophy* **69**, 347–75.
Field, H. 1973. Theory change and indeterminacy of reference, *The Journal of Philosophy* **70**, 462–81.
Friedman, M. 1980. *Foundations of Space–Time Theories*. Princeton.
Glymour, C. 1977. The epistemology of geometry, *Noûs* **11**, 227–51.
Grandy, R. E. 1973. Reference, meaning and belief, *The Journal of Philosophy* **70**, 439–52.
Nagel, T. 1979. *Mortal Questions*. Cambridge.
Quine, W. V. O. 1975. On empirically equivalent systems of the world, *Erkenntnis* **9**, 313–28.
Rescher, N. 1977. *Methodological Pragmatism*. Oxford.
Rorty, R. 1976. Realism and reference, *The Monist* **59**, 321–40.
Rosenkrantz, R. D. 1977. *Inference, Method and Decision*. Dordrecht.
Sellars, W. 1963. *Science, Perception and Reality*. London.
Sklar, L. 1976. Thermodynamics, statistical mechanics and the complexity of reductions. In *Boston Studies in the Philosophy of Science Vol. 32*, eds. R. S. Cohen *et al.*, pp. 15–52. Dordrecht.

---

[10] I thank Chris Hookway and Richard Healey for constructive comments on a draft of this paper.

Thompson, M. 1978. Peirce's verificationist realism, *The Review of Metaphysics* **32**, 74–98.

Van Fraassen. 1976. To save the phenomena, *The Journal of Philosophy* **73**, 623–32.

Williams, B. A. O. 1978. *Descartes: the Project of Pure Enquiry*. Harmondsworth.

Yoshida, R. M. 1977. *Reduction in the Physical Sciences*. Halifax, Nova Scotia.

3   *Science and the organization of belief*

S. KÖRNER

For this function [of a scientific law] is just exactly that of organizing our empirical knowledge so as to give us both intellectual satisfaction and power to predict the unknown (R. B. Braithwaite 1953: 339)

Among Richard Braithwaite's many important contributions to the philosophy of science are a clear distinction between scientific prediction and scientific explanation and an analysis of the manner in which both the predictive and the explanatory functions of a scientific theory depend on its internal structure. The purpose of the present essay is to extend the scope of this analysis by considering the dependence of the predictive and explanatory functions of a scientific theory not only on its internal structure, but also on its context in the system of beliefs of which it is a part. This extended analysis throws, it seems to me, further light on the nature of scientific prediction and explanation, *e.g.* on scientific controversies in which the opposing parties agree about the ranking of competing theories according to their predictive, but not also according to their explanatory, power.

Part I contains a sketch of the connection between the predictive power of scientific theories and their special concentration on certain features of experience to the exclusion of others. It indicates how this concentration is achieved by modifying the formal structure, of non-specialized, extratheoretical thinking, its conceptual net, or both (1); how as a result of these modifications a divergence arises between the empirical and the theoretical aspects of scientific inquiry, and how this gap is bridged in the application of theories (2). It concludes with a brief discussion of the manner in which the predictive effectiveness of theories is tested and improved (3). Part II exhibits in outline the connection between the explanatory power

43

of a scientific theory and the general organization of the system of beliefs of which it forms part. It begins by considering various types of distinction between particulars and attributes, between consistent and inconsistent beliefs, and between subjectivity and intersubjectivity (**4**). It then considers the epistemic stratification of beliefs and the structure of categorial frameworks (**5**). It concludes by showing the relevance of these aspects of the organization of an overall system of beliefs to the explanatory power of its component scientific theories (**6**).

I

1 *On the logico-mathematical structure and conceptual net of theories.* Theories differ from extratheoretical, "ordinary" or "commonsense" beliefs by their logico-mathematical structure, their conceptual net or both. The differences in logico-mathematical structure are most obvious in the case of mathematically formulated theories, especially those which employ a fairly extensive mathematical apparatus. But even theories involving a minimum of quantitative reasoning show some logico-mathematical features which are absent from "ordinary" ways of thinking. The differences between the conceptual net of theoretical and of ordinary beliefs may also be more or less obvious. Theories with an altogether different conceptual net (*e.g.* an economic and a physical theory) may have the same logico-mathematical structure, as determined by the logic and mathematics employed in their formulation (*e.g.* higher predicate logic and the classical theory of real numbers). To assert that theories *qua* theories differ from their surrounding non-theoretical beliefs is not to deny that the differences vary from theory to theory or that what at one time is a feature of one or more theories may at a later time become a feature of ordinary thinking.

In order to exemplify the difference in logical structure between theoretical and extratheoretical thinking it will be sufficient to consider two principles which are characteristic of a wide range of theories, but are normally not implicit in extratheoretical thinking, namely the principle of the exactness of concepts and the principle of the transitivity of quantitative equality. A strong version of the former principle, which implies the law of the excluded middle, has been formulated by Frege as the requirement that "the definition of a concept (a possible predicate)" must "for every object determine

unambiguously whether it does or does not fall under the concept"
(1903: 69). A weaker version, which does not imply the law of
excluded middle, merely requires that the definition of a concept
must for every object exclude the possibility that the object is with
equal correctness judged to fall under the concept or not to fall
under it. The weak principle of exactness thus proscribes concepts
with border-line cases, *i.e.* objects to which the concept is with
equal correctness attributed or refused. It does not merely proscribe
the joint attribution and refusal of a concept to an object, which is
always incorrect. (In a similar way a border-line candidate for
admission to, and hence for rejection from, a certain occupation or
rank may correctly be admitted and correctly be rejected, but
cannot correctly be admitted and rejected.) Both Wittgenstein,
together with other defenders of ordinary language against the
artificialities of specialist theories, and Frege, together with other
defenders of exact theories against the impressions of ordinary
language and its inexact "concept-like formations" (1903: 71),
agree that conformity to the strong – and thus *a fortiori* to the weak
version – of the principle of exactness distinguish ordinary from
theoretical thinking. However, the prevalence in the latter of the
principle of exactness does not exclude the existence or, at least,
the possibility of theories embedded in a logic admitting inexact
concepts.

The principle of the transitivity of quantitative equality may also
be called Poincaré's principle of quantitative equality since it was
Poincaré who clearly pointed out that (i) perceptual, empirical or
operational indistinguishability in respect of any empirical feature
is, because of the limited discriminatory range of our senses or
measuring instruments, non-transitive; that (ii) mathematical
equality or – what comes to the same – any exact concept of equality
is a transitive relation, and that (iii) whenever one replaces 'indistinguishable in a certain respect through physical measurement of a
certain type' by 'mathematically equal in a certain respect' one
substitutes a transitive for a non-transitive relation. The principles
of exactness and of the transitivity of equality are clearly not the
only ones by conforming to which theoretical thinking differs from
ordinary thinking. Thus a fundamental difference between the
physical theories and ordinary thinking about the physical world
lies in the contrast between the structure of mathematical continua,
as characterized *e.g.* by Dedekind, and the structure of empirical

continua as – more or less adequately – described by Aristotle in his *Physics* or, much later, by Brentano.

Theoretical and extratheoretical thinking exhibit not only differences in their logico-mathematical structure, but also in their conceptual nets. The opposition between the two types of difference is not always clear and depends on where one decides to draw the line between what are and are not principles of logic and mathematics. Yet, one can easily think of a theory which differs from its extratheoretical surroundings not in its logico-mathematical structure, but in its conceptual net; more precisely, in the deductive relations between its concepts. Such a difference can be the result of "theoretical abstraction", *i.e.* of removing certain concepts, together with their deductive relations to other concepts, from the extratheoretical conceptual net; or of 'theoretical innovation', *i.e.* the introduction of new concepts and corresponding deductive relations into the net.

Thus the concept of a particle in classical mechanics can be regarded partly as the result of deductive abstraction (*e.g.* of certain features, such as colour, from the extratheoretical concept of a physical object), and partly as the result of theoretical innovation (*e.g.* by characterizing the particle as being subject to the three laws of motion). Another example of a difference between the conceptual nets of ordinary and theoretical thinking is the concept of economic man in classical economic theory and the concept of man in ordinary thinking about human conduct. A fuller understanding of the difference in logico-mathematical structure and conceptual net between ordinary and theoretical thinking would require a more systematic inquiry and a wider variety of examples (*cf.* my 1966).

2 *On the divergence of intratheoretical and extratheoretical concepts and their limited identifiability.* Many philosophers of science have distinguished between the "theoretical concepts" of a theory and its "empirical concepts". They hold that the former belong only to the theory and are not instantiated in perception, whereas the latter are instantiated in perception and are common to the theory and to extratheoretical commonsense thinking. This particular distinction has to be rejected, because commonsense and theoretical thinking differ in their logico-mathematical structure or their conceptual net (or both), and while commonsense concepts are instantiated in perception, their theoretical modifications are not so instantiated. Instead one can distinguish between intratheoretical and extratheoretical concepts, and replace the customary distinction between theoretical and empirical concepts by a more adequate distinction between two kinds of intratheoretical concepts, of which one is more directly linked to perception than the other. For the sake of clear examples it is instructive to recall that many – though not necessarily all – theories conform to the principle of exactness and to compare exact theoretical with inexact perceptual concepts.

Since perceptual attributes are (with the possible exception of so-called determinables such as 'being coloured' or 'being extended') inexact, *i.e.* admit of border-line cases, no intratheoretical concept belonging to a theory which conforms to the principle of exactness is strictly speaking instantiated in perception. It was partly for this reason that Plato distinguished between the world of opinion which "tumbles about between being and non-being" and the unchanging, truly real world of the Forms and that he regarded perceptual objects as "participating in" (as opposed to instantiating) the Forms. Even if one rejects Plato's absolutism, one may find his view of the relation between perception and non-perceptual knowledge relevant to an understanding of the relation between perception and theoretical concepts. Consider, for example, a perceptual triangular object (briefly, a certain perceptual triangle), the inexact concept of a perceptual triangle (*i.e.* 'x is a perceptual triangle') and the exact concept of a Euclidean triangle (*i.e.* 'x is a Euclidean triangle'). The perceptual triangle is not an instance of 'x is a Euclidean triangle', since *e.g.* Euclidean, unlike perceptual, triangles are bounded by one-dimensional lines and, hence, invisible. It is an instance of 'x is a perceptual triangle' and it "participates" in or is linked to 'x is a Euclidean triangle'.

The link can be conceived as mediated by 'x is a perceptual triangle'. For since 'x is a perceptual triangle' and 'x is a Euclidean triangle' can in certain contexts and for certain purposes – namely whenever Euclidean geometry is successfully applied to perception – be treated *as if* they were identical, the perceptual triangles can in these contexts and for these purposes be treated *as if* they were instances of 'x is a Euclidean triangle'. Or, to use Platonic terminology, the perceptual triangle participates in the concept of Euclidean triangle in so far as this concept can be identified with (is an idealization of) the concept of a perceptual triangle. In order to avoid distracting issues of Platonic exegesis, I shall say that an intratheoretical concept which is in certain contexts and for certain purposes identifiable with a perceptual concept is (in these contexts and these purposes) perceptually linked. An intratheoretical concept which is not (in any context in which, and for any purpose for which, the theory is employed) identifiable with a perceptual concept will accordingly be called perceptually unlinked.

In this connection it should be recalled once more that there is in principle no reason why intratheoretical concepts should not be inexact and differ in other respects from extratheoretical concepts. An example would be a biological concept of species, which though admitting of border-line cases, might form part of a genetic theory the conceptual net of which is very far removed from "ordinary" thinking. Yet, whatever the difference between the intratheoretical concepts of a certain theory and the extratheoretical concepts with which some of them are conditionally identifiable, the twofold division of a theory's concepts into theoretical concepts, which are peculiar to the theory, and empirical concepts which are not, must be replaced by a threefold division of concepts, namely (I) (a) perceptually linked intratheoretical concepts; (b) perceptually unlinked intratheoretical concepts and (II) extratheoretical concepts, in particular those perceptual concepts with which the perceptually linked intratheoretical are conditionally identifiable. The conflation of the perceptually linked intratheoretical concepts (the particulars and the statements of a theory) with the perceptual concepts (the particulars and the statements with which they are identifiable) hides, or at least seriously obscures, the nature of the limited identifiability and its relevance to the predictive and explanatory function of the theory.

An important purpose of idealizing the logico-mathematical structure or the conceptual net of non-specialist ordinary thought into a theory, is the creation of predictively convenient linkages between its perceptually linked intratheoretical concepts on the one hand and perceptual concepts on the other. The predictive convenience depends to a large degree on the comprehensiveness of the idealized perceptual (or perceptually realizable) domain as well as on the clear demarcation of its limits. Examples of such domains are the domain of physical phenomena in so far as it is free from other than experimental human interference, and the domain of prudential economic conduct in so far as it is free from any incompatibility with the agents' moral convictions. The wide scope of these domains and an exclusive concern with them may absolve a scientist from the requirement of being ever mindful of the gap between theory and perception and of the limited links between them. The philosopher of science has no such licence.

3 *On testing hypotheses and improving the predictive power of theories.* A fairly simple type of hypotheses, belonging to a theory, can be formalized by means of the lower predicate logic as implications of form $(Vx)(P_o(x) \rightarrow Q_o(x))$, where $P_o(x)$ and $Q_o(x)$ are constant predicates. Such examples will serve for the moment, if the following provisos are kept in mind: Even hypotheses for the formalization of which the lower predicate logic is sufficient may be made much more complex by containing a greater number of logical constants, predicate constants and quantifiers. Many scientific theories, especially those which employ real numbers, can only be formalized within a system which allows for predicate variables and their quantification. In such systems one may formalize some or all hypotheses as statements about kinds of predicates – a procedure which has the advantage of focusing attention on the problem of the relation between a theory and its domain of application. (See *e.g.* W. Stegmüller 1973: 12f and *passim*.) Lastly, it is obvious that most scientific theories are not completely formalized, and many are not sufficiently definite to allow such formalization.

In accordance with the preceding discussion of the structure of scientific theories and their relation to what is, or can be, perceptually given, we may distinguish between three types of hypotheses, namely (i) intratheoretical formal hypotheses; (ii) intratheoretical substantive hypotheses and (iii) identifiability or linkage hypotheses. The intratheoretical formal hypotheses are the principles and theorems of the logic and of the mathematical systems incorporated into the theory. The intratheoretical substantive hypotheses are (a) the perceptually linked statements of the theory, *i.e.* statements containing only perceptually linked predicates; (b) the perceptually unlinked statements of the theory, containing only perceptually unlinked predicates; and (c) the mixed statements containing both perceptually linked and perceptually unlinked predicates. The identifiability or linkage hypotheses state the context in which, and the purpose for which, perceptually linked predicates of the theory are identifiable with theoretical perceptual predicates.

To test a theory is to test the conjunction of its hypotheses. Thus, if a predictive consequence of the theory is not borne out by accepted results of experiment or observation, the conjunction of hypotheses has to be replaced by another which can be reconciled with the available experimental and observational evidence. As Poincaré and Duhem have shown – and as in the case of the more complex theories is almost generally admitted – it is possible to replace a conjunction which in the light of certain tests has in some respects been found defective, by more than one conjunction which is free from this defect. According to what is sometimes called the "holistic" view of scientific theories, the choice among the non-defective conjunctions and, hence, the decision as to which hypotheses, if any, should be preserved and which hypotheses rejected, is a matter of taste or instinct rather than of principle. Braithwaite expresses his version of holism with characteristic clarity and acknowledges his indebtedness to Campbell and Ramsey with characteristic generosity (1953: 99). A similar version of holism is due to Sneed and Stegmüller (1973).

From what has been said about the logical structure and the conceptual net of theories, it follows that the mentioned versions – and some others which are less clearly formulated – are in one sense not holistic enough and in another too holistic. They are not holistic enough in that they do not consider all kinds of hypotheses which are subject to possible revision. Thus most holistic philosophers of science acknowledge only substantive intratheoretical hypotheses, namely perceptually linked hypotheses, which they usually call "empirical"; perceptually unlinked hypotheses which Braithwaite (1953) calls "Campbellian"; and mixed hypotheses. They rarely pay attention to formal intratheoretical hypotheses, especially to the possibility that the intratheoretical logic of a theory may differ from the logic underlying the extratheoretical ordinary thinking to which the theory is linked.

A kind of hypothesis which, as far as I can see, is generally ignored by holistic philosophers of science, are linkage or identifiability hypotheses. The main reason for this oversight is, of course, their failure to distinguish perceptually linked intratheoretical concepts from perceptual extratheoretical concepts – a failure which, as has been mentioned earlier, manifests itself in their use of the term 'empirical' to cover both types of concepts. Indeed, if the two kinds of concepts are not distinguished, the problem of their identifiability in certain contexts and for certain purposes does not even arise. Another reason why philosophers of science tend to ignore identifiability hypotheses is the view that perceptual concepts (*e.g.* '*x* is a perceptual triangle'), or their instances, are "approximations" of perceptually linked concepts (*e.g.* '*x* is a Euclidean triangle') or their instances, in a sense of 'approximation' which is considered obvious. For this reason the conditional identifiability of perceptually linked and perceptual concepts or of their instances tends to be confused with the relation between a spread of operationally measured values and a statistically determined "true" value of a quantity. Yet this relation is, at most, one species of the identifiability relation between perceptually linked and perceptual concepts.

The holistic doctrine in its usual versions is too holistic in the sense that it ignores the possibility of ranking various hypotheses and, hence, conjunctions of them according to their explanatory value. Such ranking is not simply dependent on conjectures about the comparative predictive power of different hypotheses, since it may apply even in cases where the predictive power of two competing hypotheses is considered equal. An example would be Einstein's preference for preserving the hypothesis of the principle of causality against any competing alternative on the ground that he finds non-causal theories less intelligible than causal ones, or – what is not quite the same – that he derives more intellectual satisfaction from causal than from non-causal theories. In order to clarify the notions of the explanatory power of a theory and of the intellectual satisfaction derived from it, it is necessary to inquire further into the relation between theories and their extratheoretical surroundings.

II

Philosophers of science who distinguish between the predictive and the explanatory power of scientific theories and who rank scientific theories, especially predictively equivalent ones, according to their explanatory power, tend to compare them in respect of their deductive organization, in respect of their conformity to a paradigm, or in both respects. Thus the deductivist comparison appeals to a hierarchy of premisses and conclusions in the sense that, as Braithwaite puts it, "to explain a law" is "to incorporate it in an established system in which it is deducible from higher level laws" (1953: 347). Against this the paradigmaticist comparison appeals either to already accepted theories as proper examples for acceptable ones or else to a non-theoretical apprehension of the world to which theories are *somehow* more or less adequate. Both these accounts pay too little, if any, attention to the delicate links between intratheoretical and extratheoretical thought and, hence, to the possibility of structural similarities and dissimilarities between them. The deductivist comparison, though precise and clear, is too narrowly based. The paradigmaticist comparison, though as wide-ranging as one pleases, is too imprecise and obscure. Instead of supporting these sweeping statements by a detailed criticism of various versions of the two accounts of scientific explanation, it seems preferable to develop a third account which might be regarded on the one hand as an attempt at extending the deductivist account and on the other as an attempt at making the paradigmaticist account more precise.

4   *On the cognitive organization of beliefs: particulars and attributes, logical consistency and inconsistency, subjectivity and intersubjectivity.* Before considering some other structural features which theoretical and extratheoretical thought may or may not have in common, a few words must be said about the origin of logic and the possibility of alternative logics – a possibility which has so far been taken for granted. For this purpose attention must be drawn to the anthropological fact that not only all human beings who employ scientific theories, but also almost all others, differentiate in one way or another between particulars and attributes and make judgements to the effect that some of the discerned attributes are applicable or are inapplicable to some of the discerned particulars (or ordered sets of particulars). The differentiation of experience into particulars and attributes or – as Frege put it – objects and concepts may take a variety of different forms. Yet whoever judges particulars to have or not to have certain attributes, *ipso facto* attempts to distinguish judgements which are true from judgements which are not. And in attempting this distinction, he *ipso facto* assumes that not all judgements are true and that, consequently, not every attribute is applicable to every particular. This assumption implies acceptance of what may be called the "weak" (more accurately the "weakest") principle of non-contradiction: There is a finite class of attributes and a finite class of particulars such that not every attribute belonging to the former class is both applicable and inapplicable to every particular (ordered set of particulars) belonging to the latter class.

The weak principle of non-contradiction implies a minimal condition of consistency and, hence, of inconsistency, logical implication and logical independence. It admits the possibility of four kinds of atomic statements, namely (I) true statements, *i.e.* statements in which an attribute is correctly applied (to a particular or ordered set of particulars), and would not be correctly refused; (II) untrue statements, *i.e.* statements in which an attribute is correctly refused and would not be correctly applied; (III) neutral statements, in which an attribute is correctly applied (refused) and would be correctly refused (applied); (IV) indeterminate statements, in which an attribute is neither correctly applied nor correctly refused. The weak principle of non-contradiction requires that the first two classes be not empty. As regards the others one may, *e.g.* by Frege's principle of exactness or other principles, enforce their emptiness. Again one may by further postulates so determine the nature of untrue statements that one arrives at a two- or many-valued logic. Having – explicitly or implicitly – chosen one of these options, a variety of other possibilities have to be considered, *e.g.* the admission of only finite, or of finite and infinite, domains of particulars. In all these respects the logic of a theory may differ from the logic of the extratheoretical thinking to which it is linked.

Another aspect of the organization of intratheoretical and extratheoretical beliefs, which may be the source of pervasive differences between the conceptual nets of a theory and extratheoretical thinking, is the way, or more precisely the variety of possible ways, in which intersubjectivity is conferred on subjective phenomena. That an object, say o, which is given in my perception, is also intersubjectively given is not a perceptual feature of the object. More precisely, in judging that o, which has certain perceptual attributes, is also an instance of 'x is intersubjectively given', one is not applying a further perceptual attribute to o but an *a priori* concept in Kant's sense, *i.e.* an attribute which is not abstracted from what is given in perception, but is nevertheless applicable to what is so given. Normally, intersubjectivity is not simply conferred by the application of 'x is intersubjective', but by one or more *a priori* concepts – *e.g.* the Kantian Categories of substance, cause and interaction – the applicability of which logically implies the applicability of 'x is intersubjective'.

Whereas Kant held that intersubjectivity can be conferred on subjective phenomena in one way only, namely through the application of the Categories, whose uniqueness he believed himself to have demonstrated by a "transcendental deduction", post-Kantian science involves in some of its branches the application of non-Kantian intersubjectivity-concepts. Thus the substance which is conserved according to the theory of relativity differs from the substance as conceived by Newton, and the causally deterministic laws of Newtonian physics differ from the probabilistic laws of quantum mechanics. (For a more detailed discussion of intersubjectivity-concepts see my 1978.) If, as seems fair, one assumes that a great deal of contemporary extratheoretical thinking employs the Kantian Categories, whereas some theories linked to it involve a different set of intersubjectivity-concepts, one is provided with a good example of a pervasive difference in the conceptual nets of a theory and the extratheoretical thinking to which it is linked.

5   *On the epistemic stratification of beliefs, supreme beliefs and categorial frameworks.* A person's logical principles determine which of his beliefs, if any, are incompatible with each other. The acceptance of these principles requires that if he becomes aware of an incompatibility between two internally consistent beliefs he reject one of them. The requirement does not mean that the decision as to which of the beliefs is to be rejected has to be made immediately. Nor – as we have seen when comparing intratheoretical and extratheoretical beliefs – does it mean that the rejected belief cannot be preserved in a modified form. Lastly, the requirement that, of two incompatible beliefs, one has to be rejected does not determine which of them has to be rejected.

The decision as to which of two incompatible beliefs is to be rejected need not be *ad hoc*, since the beliefs of a person or group of persons are normally stratified in the sense that some beliefs take precedence over others. Examples are a Catholic's beliefs in the bible as compared with his historical beliefs, or an empirical scientist's experimentally grounded beliefs as compared with his empirical generalizations. The relation of epistemic precedence or domination can be defined as follows: A class of a person's beliefs, say α, dominates another class of his beliefs, say β, if and only if, in case he discovers that an internally consistent belief belonging to α is inconsistent with an internally consistent belief belonging to β, he rejects the latter. In line with this definition, a class of beliefs α of a person is the class of his supreme beliefs if, and only if (i) α cannot be decomposed into two classes α_1 and α_2 such that α_1 dominates α_2 and (ii) α dominates β where β is the class of all those beliefs of the person which do not belong to α. The class of a person's supreme beliefs includes the principles of his logic (or, if he employs more than one system of logic, the logical principles belonging to α).

The definitions are meant to allow for different types of domination, ranging from a domination of a person's beliefs belonging to α over his beliefs belonging to β which is hardly distinguishable from a convenient habit, to his explicit acknowledgement of a superiority of α over β. Such a superiority may rest on his submission to a secular or spiritual authority, on prudential considerations, on an aesthetic vision of the world, or it may have still other grounds. The membership of the dominating and the dominated classes – in particular of a person's class of supreme beliefs and its complement – may also vary greatly. Thus the class α of his supreme beliefs may consist of, or contain, beliefs about particular events, *e.g.* those related in the New Testament. It may consist of, or contain, empirical generalizations, principles about the manner in which objectivity is conferred on subjective phenomena, formal principles, *etc.*

The systems of supreme principles which are particularly relevant here are systems (more precisely, systems or subsystems) of supreme principles which have developed in our culture and have in one way or another taken account of its logical, mathematical and scientific theories. The clearest formulations of such systems are due to philosophers who, from early Greek times to the present day, have tried to make accepted principles explicit, or to propose principles, constraining all extratheoretical and theoretical thinking. The principles are intended as answers to questions as to what kinds of intersubjective entities exist, how they are constituted and individuated, and how the consistency of thinking about them is to be safeguarded.

What has been said about these issues can for our purposes be summarized by defining the notion of a categorial framework as a set of supreme beliefs which determine:

    (i) a differentiation of experience into particulars and attributes and, hence, a logic containing at least the weak principle of contradiction without which the differentiation would be empty;

(ii) within the constraints of this differentiation and logic a categorization of all intersubjective particulars into maximal kinds (*maxima genera*) and, hence, the assumption of the non-emptiness of certain *a priori*, intersubjectivity-concepts whose application to subjective phenomena interprets them as intersubjective;

(iii) (a) for each maximal kind a constitutive principle according to which an object being a member of the maximal kind logically implies its possession of a certain attribute or conjunction of attributes;

(b) for some maximal kinds, if any, an individuating principle according to which a member's distinctness from every other member of a maximal kind logically implies its possession of a certain attribute or conjunction of attributes (*e.g.* the specific spatio-temporal position of material objects in the Kantian ontology);

(iv) principles for distinguishing between ultimate particulars and dependent particulars which can be analysed in terms of ultimate ones.

Although this definition seems sufficiently specific for our purpose, it may well turn out that some further specification would add to its usefulness. Again, a person's adherence to a categorial framework may – like his adherence to a morality, a legal system, a scientific theory and other systems of beliefs and attitudes – be more or less explicit, more or less definite and more or less confident. It also may, and frequently does, happen that a person is wavering between one or more systems. All this is no objection to the use of the notion of a categorial framework, since it is in general more useful to know than not to know that about which one is not certain or undecided. It should be fairly obvious that the notion of a categorial framework relativizes some traditional philosophical theses (*cf.* my 1970, 1974).

6 *On scientific theories as explanations.* The explanatory function of a theory is obviously connected with the deductive organization and epistemic stratification of its statements, in particular its axioms or laws and its testable generalizations. Consider, for example, the typical case of a theory, the axioms of which logically imply lower level hypotheses which in turn logically imply certain testable generalizations. If the testable generalizations are well confirmed by experiments or observations, it may make good sense to say that – *in so far as they are considered in isolation from their extratheoretical context* – the laws explain the lower level hypotheses and, hence, the testable generalizations.

Yet, because a theory is, through *as if* identifications, linked to extratheoretical perceptual statements, its explanatory power depends also on the relation between, on the one hand, its logic-mathematical structure and conceptual net and, on the other, the logical structure and conceptual net of extratheoretical thinking, in particular the supreme principles constituting its categorial framework. It may, for example, well be that certain general statements which logically follow from a given theory, and which in virtue of their *as if* identifiability with perceptual statements have high predictive power, may yet be considered as devoid of explanatory power because they grossly misrepresent the perceptual statements and the categorial framework in which they are asserted. It is for such reasons that Leibniz, while acknowledging the predictive power of Newton's "Democritean" physics finds it wanting in explanatory value, and why Einstein, while granting the predictive efficacy of classical quantum mechanics, continued to search for a causal theory which would make the phenomena covered by quantum mechanics not only predictable, but also intelligible.

A few definitions will help in making the point more clearly: Let us call a statement belonging to a theory a simple state-description if, and only if, it describes in the language of the theory the state of the whole or part of its subject matter at a certain time; and let us call a statement a predictive state-description if, and only if (i) it is the result of combining simple state-descriptions by means of logical connectives and quantification over individual variables and if (ii) it is logically equivalent to one or more statements to the effect that one simple state-description (non-logically) implies another. Since simple or predictive state-descriptions, as here defined, may or may not be perceptually linked, one must distinguish between the two kinds. In what follows the expression 'state-description' is always meant to be understood in the sense of 'perceptually linked state-description'. If t is a predictive state-description, belonging to a theory T, then t may be defined at "theoretically necessary" with respect to T if, and only if, it logically follows from T (*i.e.* its laws and logical principles); and as "predictively effective" if, and only if, by way of its being linked with an extratheoretical perceptual statement, say V, it is borne out by observations or experiments. That t be theoretically necessary and predictively effective is a necessary but not a sufficient condition of its being explanatory.

Since t is an idealization of V it may differ in a variety of ways from it. It may in particular, conform to or violate the supreme principles of the extratheoretical thought to which V belongs, especially the conjunction of supreme principles, say F, which constitute its categorial framework. In the light of the examples drawn from Leibniz's and Einstein's discussion of the explanatory value of certain scientific theories, to which similar examples could be added, it appears that if t is to be explanatory it must, in addition to being theoretically necessary and predictively effective, conform to F. In other words, we may define a state-description t belonging to T as "epistemically adequate" if, and only if, it does not violate the supreme principles F to which V (with which t is identified) is linked. And we must require that in so far as t is explanatory – as opposed to merely predictive – it must be epistemically adequate.

The epistemic inadequacy of the state-descriptions of a theory T, and hence of T, may rest on two kinds of divergence of T from F. One is a divergence in logico-mathematical structure, e.g. if the logic underlying T is exact (three-valued, etc.) and the logic underlying the extratheoretical thinking subject to F is inexact (two-valued, etc.) The divergence may be avoided or lessened by regarding the logic underlying T as merely auxiliary and by showing that, and how, it can be represented in the logic of F. A second kind of divergence is a divergence between the conceptual net of T and that employed by extratheoretical thinking subject to F, especially the divergence between the intersubjectivity concepts of F (e.g. the Kantian Categories of substance or causality) and those of T (e.g. the category of probabilistic connection in quantum mechanics). Here again the divergence can be avoided or lessened by regarding the divergent concepts of T as merely auxiliary or even as useful fictions. The divergence between T and F may, but need not, lead to a revised epistemic stratification, as a result of which previously supreme principles of F become dominated ones.

To theoretical necessity, predictive effectiveness and epistemic adequacy, as conditions of the explanatory function of a theory's state-descriptions and, thus, of the theory, there must be added certain practical and aesthetic requirements, e.g. that the deductive structure of the theory be not so loose nor its conceptual net so lacking in transparence that the employment of the theory produces intellectual discontent rather than the intellectual satisfaction which Braithwaite rightly requires. That an explanation must be the object of practical or aesthetic pro-attitudes is generally agreed, although an analysis of their specific nature cannot be undertaken here. (For a general discussion of the relation between cognitive and practical rationality see my 1976, Ch. 16.)

It should be noted that the preceding account of the conditions which together are meant to define the explanatory value of a theory does not exclude scientific progress of various kinds: It admits progress within the limits of the theory's axioms, *e.g.* by simplifying its deductive structure or by making it technically easier to increase the number of testable, predictive state-descriptions. It admits progress within the limits of a categorial framework, *e.g.* by unifying a plurality of theories into one or by replacing a theory which contains auxiliary fictions by one which is free – or more free – from such blemishes. It even admits, though it does not suggest, the possibility of a progressive sequence of categorial frameworks. However, a discussion of this possibility belongs to transcendent metaphysics and lies outside the scope of this essay.

Yale University
University of Bristol

REFERENCES

Braithwaite, R. B. 1953. *Scientific Explanation*. Cambridge.
Frege, G. 1903. *Grundgesetze der Arithmetik*, Vol. 2. Jena.
Körner, S. 1966. *Experience and Theory*. London.
Körner, S. 1970, 1974. *Categorial Frameworks*. Oxford.
Körner, S. 1976. *Experience and Conduct*. Cambridge.
Körner, S. 1978. Über ontologische Notwendigkeit und die Begründung ontologischer Prinzipien, *Neue Hefte für Philosophie* **14**, 1–18.
Stegmüller, W. 1973. *Theorie und Erfahrung*, 2. Halbband. Berlin.

4 Braithwaite and Kuhn: analogy-clusters within and without hypothetico-deductive systems in science

MARGARET MASTERMAN

1 Current relativist conceptions of science depend widely, though vaguely, upon the insights of T. S. Kuhn (1962); and, in particular, upon his notion of a paradigm. This notion is being used by relativists to support the contention that, since scientific theory is paradigm-founded, and therefore context-based, there can be no one discernible process of scientific verification. However, as I have shown in an earlier paper (1970), there is another, more exact conception of a Kuhnian paradigm to be considered; namely, that conception of it which says that it is either an analogically used artefact, or even sometimes an actual "crude analogy"; that is, an analogical figure of speech expressed in a string of words.

This alternative conception of a paradigm, far from supporting a verification-deprived conception of science (which, for those of us philosophers who are also trying to do technological science, just seems a conception of science totally divorced from scientific reality) can, on the contrary, be used to enrich and amplify the most strictly verification-based philosophy of science which is known, namely the Braithwaitean conception of it as a verifiable hypothetico-deductive (H-D) system. For such a paradigm, even though, in unselfconscious scientific thinking, it is usually a crude and concrete conceptual structure, can yet be shown to yield a set of abstract attributes. These can provide "points", or "nodes", or other more complex units, on to which some even more abstract H-D system can then, like a mathematical envelope, be "hung"; after which the power of the mathematics can "take off on its own". And, although Braithwaite's whole account of science is, I think, over-simplified, yet he is right in showing how, from that point on, the mathematics can be used. For it can indeed be used progressively to fit on to, and to test, features of a second concrete, but also verifiable and operational, B-component (the concrete analogy which started everything off being the A-component); and, even in comparatively undeveloped science, let alone in advanced science, this kind of verification really does occur. But for the Braithwaitean H-D model to be realistic, the crude analogy, the A-component, has got to be there also; because its function is either to guide, and orient, the subsequent mathematical (or mechanical) development when this development has occurred, or, predictively speaking, to do instead of it when it has not yet occurred – for theory-making comes at quite a late stage, in real science.

This point needs amplifying. In real science, as Norman Campbell, who knew about it, long ago said (1920), there is nearly always far too much mathematical "play" in any mathematics which is powerful enough to be used for scientific development. The mathematics produces infinities; it fails to produce what you want; the complexities, which were required to make it fit on to the original analogical insight in the first place, later on, when inappropriate theorems have to be brought to bear, make it altogether out of hand, so that it generates nothing which you can any longer recognize. Its excesses and rigidities have to be tailored; moreover, operationally valid fudges (which make it fit the B-component) have to be organized. Moreover again, it may become necessary to shift from one mathematical system to another, in order, predictively and in the end, to get anywhere. And by what other system of predictive ideas can the mathematically predictive system of ideas itself be tailored? And, if it is necessary to do mathematical shifting, what guides the shift? Only the original (Kuhnian) analogical insight which started the whole enterprise off in the first place; for in abstract deductive science as it is actually done, the second kind of fitting, which permits the verification, normally only comes into action right at the end.

So you cannot work backwards, in orienting mathematical development. If the whole enterprise is to be predictive – and prediction is the object of it – you have got to work forwards; and there is nothing else to work forwards from, except the original crude paradigm.

That this evident fact has not been seen by philosophers of science is due, I think, to many diverse causes. One cause is that epistemologists and logicians (from the ranks of which philosophers of science are normally recruited) have relegated the study of analogy to the English Department. Another cause is that (speaking broadly) analogy is used to form scientific paradigms in a way converse to that in which it is used in poetry – a point to which I shall return – so that nothing that is said about it in English Departments is likely to be serviceable to the philosophy of science. And the third cause was that, whereas the special skills of analytic philosophy were precise conceptual analysis and/or conceptual model-making (which in the older philosophy used to be called 'rational reconstruction') no technique was available for applying these skills to analogy; so that it would have seemed plain unintelligible to say that an analogy could form a predictive structure, on to which a mathematical or mechanical system could then subsequently be "hung".

In my view, the missing technique has now become available. And therefore, in this second paper on the nature of a Kuhnian paradigm. I propose to use the technique to make a model of analogy; thus modelling that primitive modelling activity of science which still persists, even when the vehicle for it is not a three-dimensional artefact, or even a two-dimensional schema, or "picture" or diagram, but only a one-dimensional stretch of natural language.

To make this model I shall indeed have to cross the academic disciplines, and this will cause problems, in that I will be accused of no longer doing philosophy. But I shall not be retreating into literary criticism. On the contrary, I shall be offending literary critics by manipulating natural language with an imaginary computer, and also, probably, offending philosophers by saying that to do this philosophically requires making three adjustments to the current customary conception of philosophy. (And yet, is it not high time that we widened philosophy?)

The three adjustments which I need to make are the following:

(1)   I need to extend the current sense of 'deduction' so as to be able to say 'The sentence S_2 *is computable from* the sentence S_1 in the coded language L', as a philosophic replacement for 'The formula F_2 *is deducible from* the axioms $A_1, A_2, \ldots, A_n$ in the axiomatized calculus C'. (To gain more light on why I need to make this change, see Figure 3.)

(2)   Secondly, I need to persuade philosophers of science, and notably Hesse (1974), to reverse the whole direction of approach

which they now take when they are wishing to incorporate a conception of analogy within the contemporary universe of discourse of the philosophy of science. For they, in order to remain academically "within the literature", always water down the real characteristics of language when talking of analogy, as the earlier logicians did when talking of Moore's technique (*cf.* Masterman 1961), in order to keep just so much of the phenomenon of natural language in their conception of science as will enable the phenomenon of analogy, though by free association, to be included in it also. But in fact, as I think my model conclusively shows, if you take either natural language or analogy seriously, you cannot do this.

(3)   The third adjustment is to the currently fashionable conception of the nature of language as it is referred to from within philosophy, and it consists in saying that Hacking's (1975) emasculatory trail, from the "Heyday of Ideas" through the "Heyday of Meanings" to the far more superficial, though more systematic, "Heyday of Sentences", has got to be retrod. Analogy, like metaphor, is the superimposition of one framework of *ideas* upon another; so, to analyse it, you have to have a model which, in an unashamedly seventeenth-century manner, though with a new gloss, deals in ideas. For the seventeenth-century philosophers were operationally right in their inherited belief that (in some sense) ideas "lay behind" words. But they were operationally wrong in their conviction that these same ideas which, as they themselves admitted, formed the root and basis of public language, were ascertainable only by private introspection, with no public provenance.

What has happened, in the passage from the Heyday of Ideas to the Heyday of Sentences, is that philosophers with an exaggerated reverence for mechanism have tried at all costs to find something in language to mechanize. Grammatical transformation (Chomsky), propositional connection (Russell *et al.*), verificational systematization as between fact-sentences and first-order or other predicative sentences (Tarski, Quine, Montague, Davidson), systematization of speech-acts (Austin to the future through Searle): they have all done it. What all these philosophers have forgotten, when calling the resulting systematization 'a language', was that all the rest of what was really there in language – and all that really matters about it, once you are no longer doing logic – was still being fully and efficiently processed by them themselves, intuitively, subliminally, non-consciously. But now, in the computer world of word-processing, we put real language into a real machine; and this machine really is an inert mechanism: it has no sublimen. And the result of this, of course, is that all the semantically shifting layered and interlacing depths of language – all the most Coleridge-like features of this frightening and volatile phenomenon of human talk, the very foundation of thinking – are now progressively coming out into the light.

How many already identified philosophical problems will have to be solved, using the new methods, before those who refuse to admit to their existence become castigated (probably in future works of Hacking) as 'know-nothing' philosophers (1975: 163) is anybody's guess. Here, in a first whiff from the new world, I try to apply them to analyse analogy.

2 The analogical use of language, on this model, becomes only a special case of the normal use of all language; that is, when "the normal use of all language" is interpreted operationally, not philosophically. For language, when coded and dictionary-matched on a machine, does not turn out to consist of isolated, single-meaning sentences. Real language consists of a *reiterative semantic flow*, the total sequence of the units of which combine to form a *text*. These units are, predominantly, 7 (plus-or-minus) word *phrases*, which are often coded within nested brackets as being *lists*; for on this model, the primitive unit of knowledge is an item on a list (the unit "booked" in a "book" – a book was originally a list: *Shorter Oxford English Dictionary* 3rd edn, 1977: 217), not a sentence. Each such item, or phrase, is indeed built from a sequence of stressed or unstressed *words*. But, in the general model, though not in all particular ones, every single word in every such sequence has both multiple syntactic uses (called *homographs*) and also, when considered semantically, a whole string, if not an actual whole structure, of multiple meanings, consisting of the ways in which it is used.

Thus it is the whole coded phrase, or list item, which is sometimes – though by no means always – reminiscent of a logician's *term*. A coded word, at its simplest, is a whole partially-ordered set on its own, a *fan*: a totally different animal.

1.1 The model-maker's philosophic first question is, how the fans

66

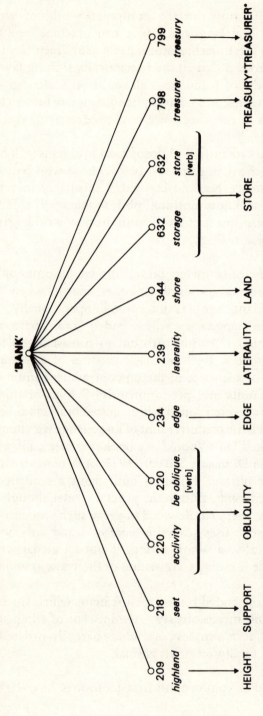

Figure 1 Fan of uses of the word 'bank', from *Roget's Thesaurus*

Notes: The labels in italics give a guide for subsequent assignment of Aspect-Markers; the Semantic Area labels themselves are in capitals. In structure, *Roget's Thesaurus* is a very rough and ready knowledge-system; on the other hand, the actual groupings of synonyms have been chosen with great care.

* TREASURY and TREASURER are, in fact, aspects of a single Semantic Area.

of word-uses are to be coded, within an imagined coded language, *L*; and the answer is that they have to be coded by mapping each of them on to a system the units of which consist of the *most-talked-of-semantic-areas* of the language. Since these areas, which are wide, are also themselves structured, to include all possible *aspects*, syntactic, referential and analogical, under which any subject of talk can be talked about, there is only one practical way of codifying them, and that is to assign to each area the name of some very general idea. Moreover, if the *area-system* is to be converted (computationally) into a *knowledge-structure*, (1) the aspect-*markers*, which recur from topic to topic, will be even more abstract than the *area-codings*; (2) to separate the analogical references from one another, the semantic areas will have to be *cross-referenced*; (3) the main overall inclusion-relations within and between the semantic areas will themselves have to be labelled with *classifying labels* so that the machine can generate *inference schemata*. And from all this it becomes evident that such a knowledge-structure, though it remains a *mappabile* of often concrete individual word-uses, will have no lack, within itself, of abstract attributions; the names of these also being word-uses, though sometimes specially constructed word-uses (for, seen fan-wise, the model bends back upon itself).

To discuss the complexities and handling-problems of such a model, and of what happens when some attribute in it goes "above the meaning-line", all this would take us way out of philosophy. The point here is that (to the extent that any selection of semantic areas which is used for processing is really taken from among the most-talked-of subjects within any language) these semantic areas, though unobservable, have objective validity. Thus, in this century, ideas reappear as the objects of content; but this time round they are publicly inferable unobservables, rather than privately introspected entities.

If we are to be simple, and thus philosophical, we must go right back in our minds, via the nineteenth-century Royal Society Library, to the seventeenth century itself and imagine ourselves re-structuring *Roget's Thesaurus* (1962) in a manner reminiscent of Bishop Wilkins' *Character Universalis* (1668). Not that simple, you will say; but now consider, using Roget, the actual fan of uses there given for the notoriously ambiguous English word 'bank' (Figure 1).

68

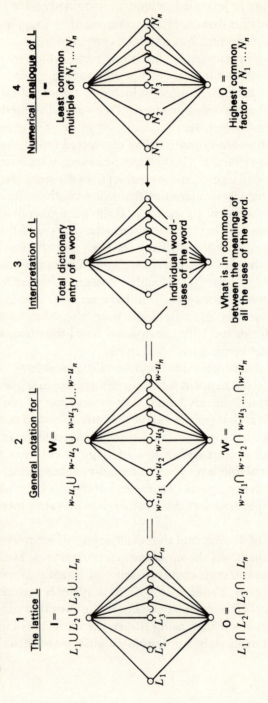

Figure 2 General schema of a dictionary entry as a lattice (Hasse diagrams)

1
The lattice L
I =
$L_1 \cup L_2 \cup L_3 \cup \dots L_n$

O =
$L_1 \cap L_2 \cap L_3 \cap \dots L_n$

2
General notation for L
W =
$w\text{-}u_1 \cup w\text{-}u_2 \cup w\text{-}u_3 \cup \dots w\text{-}u_n$

'W' =
$w\text{-}u_1 \cap w\text{-}u_2 \cap w\text{-}u_3 \dots \cap |w\text{-}u_n$

3
Interpretation of L
Total dictionary
entry of a word

Individual word-
uses of the word.

What is in common
between the meanings of
all the uses of the word.

4
Numerical analogue of L
I =
Least common
multiple of $N_1 \dots N_n$

O =
Highest common
factor of $N_1 \dots N_n$

Notes: In 1, the lattice L interrelates the unordered set of points $L_1, \dots L_n$.
In 2, W is the total fan of word-uses of a word; $w\text{-}u_i$ is a word-use of the word; 'W' is the property of being designated by the word.
In 4, N_i is any factorizable number.

In this figure, the word 'bank' is at the apex, and the names and numbers of the subjects-which-it-is-most-used-to-talk-about (call them *heads*, though they are actually neo-Wittgensteinian 'families') separate out the uses of the word and so form the spokes of the fan. If you turn the figure upside down, then a whole sequence of text is produced consisting only of a string of overt ideas, with the word 'bank' then becoming the unobservable connective between them. To enable this inversion to be performed, when required, it is advisable to extend the fan into a *lattice*, of which the I-element will be the total dictionary entry of the word 'bank', and the O-element anything which may be in common between its uses. Minimally, this common element will be the bare fact that all these subjects-of-discourse, in English, are referable to by making use of the same word 'bank'; but sometimes also there is some discernible common element of meaning between all of them or some of them, as there is, for instance, between the three uses of 'bank' which refer to *store*, *treasurer*, and *treasury*. The general scheme of such a lattice is given in Figure 2.

1.2 The question now arises as to how to make the machine detect, for operational purposes, that meaning of 'bank' which is actually used in any given text. The human being will at once answer, 'By embedding it in a phrase': 'in the savings bank', 'up a steep bank', what have you. This answer is correct – but for the difficulty produced by the fact that 'up', 'savings' and 'steep' will all themselves have to be coded into fans of uses; and thus will all of them also be ambiguous (Figure 3).

The simplest algorithmic answer to this problem is to make the constituent fans of a phrase pare down each other by retaining only the spokes with ideas which occur in each. The application of this algorithm retains the area ACCLIVITY 220 in 'steep' and 'bank'; whereas it retains as common between 'savings' and 'bank' both of the two areas STORE 632 and TREASURY 799. Note also that the general application of this algorithm and no other produces a reiterative model of the foundations of language, the main current application of which is to coordinate indexing in libraries (Figure 4).

1.3 However, when applied to the full variegation of natural language, reiterative meaning-specification is not as simple or one-

Figure 3 Fans for uses of words in 'Up the steep bank' and 'In the savings bank'

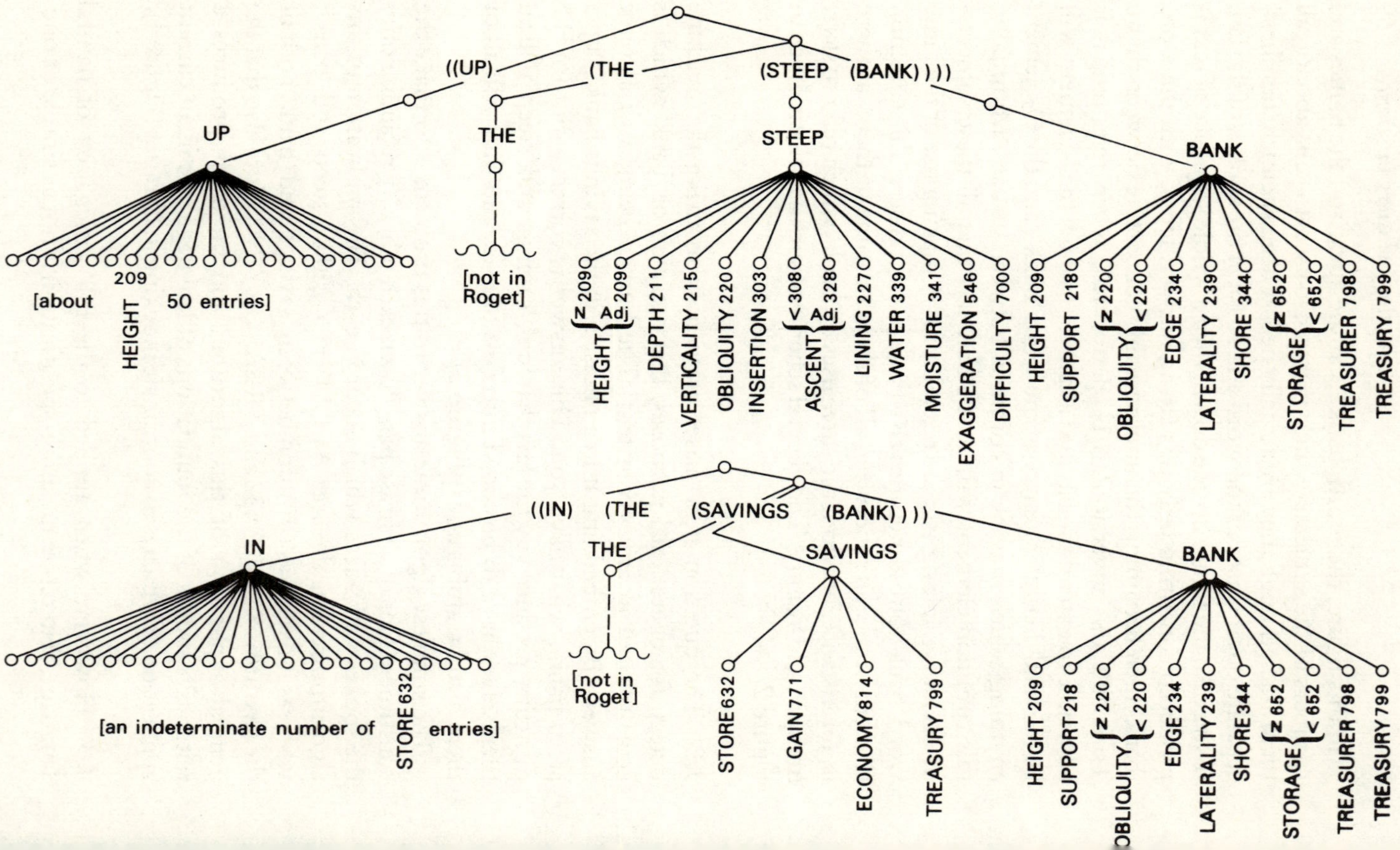

staged as this. For instance, if we had said 'grassy bank' instead of 'steep bank', we should have had to use Roget's (erratic) cross-reference system in order to get back, from 'grassy', *via* PLAIN 348, to LAND 344, which then intersects with 'bank' at SHORE. (And, in a better structured knowledge system, proceeding reincarnation-ally from Wilkins to Wilks (1975), we should operate with a layer of aspect-markers, in a manner reminiscent of the seventeenth-century Logic of Predicables, to detect that, whereas banks are regions of land which are often COVER-ABLE, grass is a plant which often acts as a COVER-ING.)

There are other cases. In bad metaphysical writing, for instance, there is far too much overlap, so that far too many spokes of all the fans remain (Figure 5).

1.4   In imaginative writing, however, in which not every sentence is a cliché, the writer can often deliberately avoid any overlap at all: 'It was an unbelievable restaurant; there were no tables, no dining-room, no waiters, no kitchen, no food; only pills and an open space in which to do physical exercises.' Here 'restaurant' really means 'anti-restaurant', though with a difference: and so there will never be a semantic overlap, whatever you do, between the fan for 'un-believable' and the fan for 'restaurant'. Here the solution is, as the reader will already have guessed, to intersect 'restaurant' with

---

Notes: In order to make this still "deductive" model simulate text (using 'deduc-tive' here in the widened sense which I mentioned earlier), I have combined the words of the text by using a bracket-nesting notation. This not only enables bracketting-patterns to help guide semantic connectives – which makes bracket-ting a much more powerful combining device than word juxtaposition – but bracketting can also be used to simulate a general emphasis pattern which later, if desired, can have syntactic annotations attached to it.

Thus two incompatible mathematical systems are both used here: finite lattice theory, which is commutative and associative; and a list-processing system derived from Church's Lambda Calculus, which is neither. Moreover, inside the machine, the whole semantic knowledge-structure will be handled as a data-base, probably within one giant module, whereas, quite elsewhere, the lists and sublists of the text will be handled by text-editing techniques, with the lot under the ultimate control of some supervisor, or control program – and all this without the predicative path through from input to output ever being lost.

In my view, the only way to think about this construction simply, and therefore philosophically, is to widen our sense of deduction so as to allow for operationally valid deductive jumps; not just to draw a picture of a black box and hang it in an ivory tower.

Figure 4   Idea-structures and their reiterative transformations

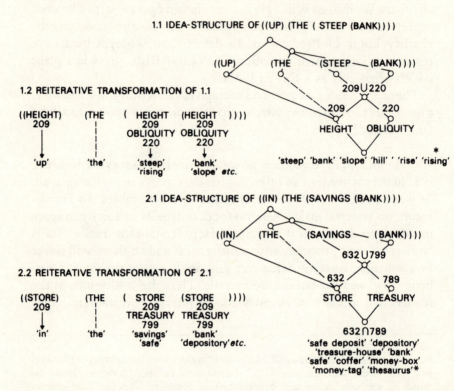

Note: * Taken from *Roget*: the intersection between two semantic areas in *Roget* is interpreted as the set of word uses which occur in both.

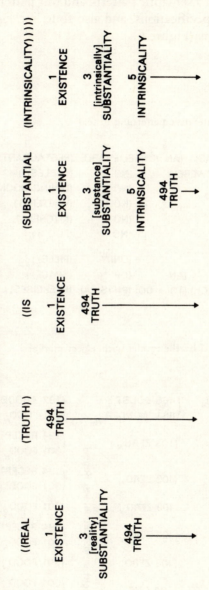

Figure 5 Excessive metaphysical overlap (exhibited by using the model)

Note: For the lists of common word uses, see *Roget.*

'dining-room', 'waiters', 'kitchen' and 'food', and 'unbelievable' with the reiterations of 'no'. But to do that the machine has to be provided with a semantic pattern; and this pattern has to have both intersection-specifications, and also four, six, eight or more numbered positions (Figure 6).

Figure 6 Reiterative patterning in text

```
((IT (WAS)) (AN (UNBELIEVABLE (RESTAURANT)));
(THERE (WERE))          (NO          (TABLES)),
                        (NO          (DINING-ROOM)),
                        (NO          (WAITERS)),
                        (NO          (KITCHEN)),
                        (NO          (FOOD));

                        (ONLY        (PILLS))
(AND)        (AN        (OPEN        (SPACE))
(IN (WHICH) (TO + DO) (PHYSICAL (EXERCISES)).)
```

(1) Text analysed by the model with aspect markers[1]

(( ))⊣	(486 DOUBT,$_N$ / 864 WONDER,$_N$)		(192 ABODE$_{H,S,L}$ / 301 FOOD$_{H,R}$))
(( ))⊣	(103 ZERO$_N$		(194 RECEPTACLE$_{H,S,L}$ / 301 FOOD$_{H,R}$))
[too many entries; see Roget]	(103 ZERO$_N$	[slight simplification: see Roget]	(194 RECEPTACLE$_{H,S,L}$ / 301 FOOD))
	(103 ZERO$_N$		(301 FOOD$_{H,R}$))
	(103 ZERO$_N$		(194 RECEPTACLE$_{H,S,L}$ / 301 FOOD$_{H,R}$))
	(103 ZERO$_N$		(301 FOOD$_{H,R}$))
	(88 ONE$_N$		(301 FOOD$_{H,R}$ / 658 MEDICINE$_{H,R}$))
(( ))	(201 INTERVAL$_{Sp,L}$		(201 INTERVAL$_{Sp,L}$))
(( )) (+)⊢	(534 EDUCATION$_{H,⊢}$)		(534 EDUCATION$_{H,⊢}$))

(2) Position-and-flow pattern of text, giving intersection specification

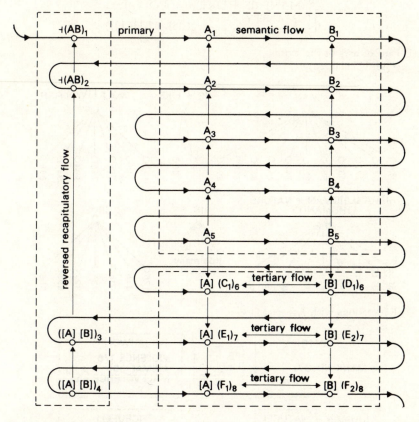

(3) Reiterative pattern-schema

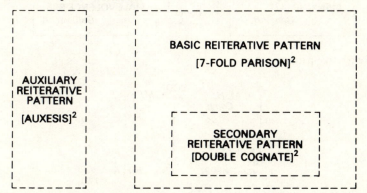

Notes: *Aspect Glossary*: N = Nullifying Aspect; H, S, L = Human Social Location; H,R = Human Renewal; Sp,L = Spatial Location; $H, \vdash$ = Human Action; $\dashv$ = Assertion of Existance; $\vdash$ = Assertion of Action.

Positional Key: A = $_N(486\cup860\cup103)$; B = $_{H, R}\cup_{H, S, L}(192\cup194\cup(301))$; C = $_N(88\cup(486\cup860\cup103))$; D = $_{H, R}(658\cup301)$; E = $_{Sp, L}(201)$; F = $_{H, \vdash}(534)$

[1] If the Aspect-Markers reiterate, you can make a join of the Semantic Areas even if these do not.

[2] The names in square brackets in (3) are names of classical Figures of Rhetoric.

Figure 7   An Analogy

$$\text{(MAN) ((IS + LIKE) ((A(WOLF))))}$$
$$\binom{\text{HUMANITY}}{\text{HUMAN + NATURE}}\text{((IS) (CRUEL)))}$$

(1) Fans of uses of the words
    (slightly enlarged)

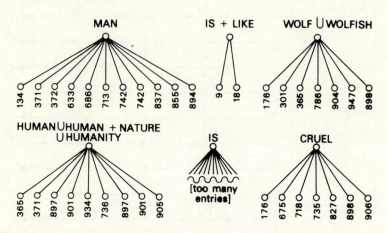

(2) Reiterations with Aspect-Markers

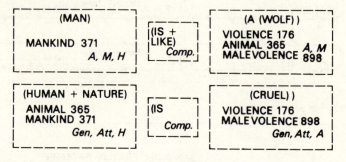

(3) Reiterative pattern

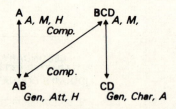

Notes: The example is after Black 1962.
   Aspect Glossary: *A* = Animate; *Att.* = Attribute; *Char.* = Characteristic; *Comp.* = Comparison; *Gen.* = General; *H* = Human; *M* = Mammal.
   Positional Key: A = MANKIND; B = ANIMAL; C = VIOLENCE; D = MALEVOLENCE

This last device, of employing a reiterative semantic pattern and then filling in the holes in it, is what we use when we model analogy. For when we draw an analogy, we are proposing to open up a new much-to-be-talked-about semantic area, rather than just drawing upon old ones; and this we can only do by filling in already known positions, in known patterns, in new ways (Figure 7).

Here, using Roget, 'man' by no means directly intersects with 'wolf'. A "man", for Roget, was by no means a wolf: far from it. Nevertheless, when we begin to fill in the pattern, connections begin to display themselves which then allow the patterning to "force the analogy" (Figure 8).

Figure 8 The progress to analogical inference

Analogical reiteration pattern
(with Aspect-Markers as in Fig. 6)

Analogical inference schema
(with Classifying Labels)

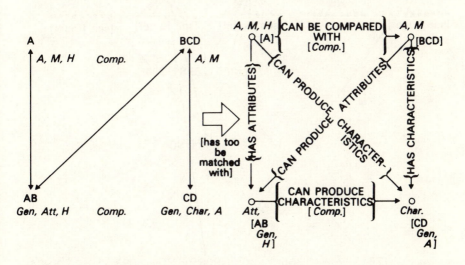

Notes: Whereas the Aspect-Markers refer to groupings of word-uses in *Roget* (and thus reinforce the cross-reference system), the Classifying Labels specify different types of inclusion-relations embedded in schemata which, when matched with appropriate reiterative patterns, enable further statements to be elicited which are not contained in the text. See, further, Figure 10.

2.1 Since analogy-drawing, on this model, and as I earlier said, is only one very normal instance of the way in which all language works, the two ways to develop analogy are, semantically speaking, the same two basic ways in which all semantic flow develops down a page. And note that, in each case, we finish up with a structure.

Figure 9 Extending an analogy: Idea structure of an analogical pile-up

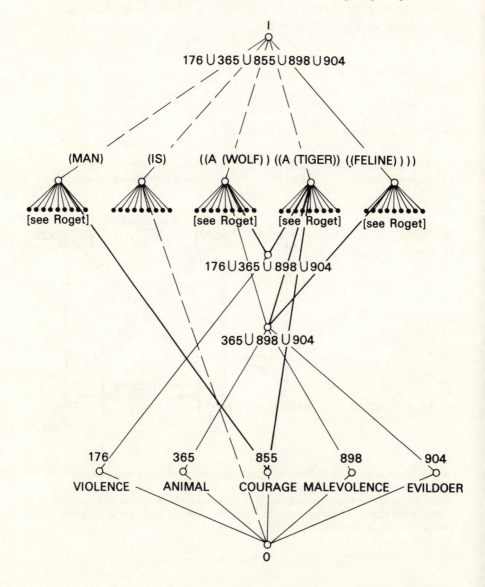

The first way to extend analogy is by analogical pile-up, the result of which is to form an *analogy-cluster*. This is done by simply drawing on more synonyms, from Roget, to extend your text; an operation which, in principle, is easy to mechanize, since Roget's semantic areas are themselves, predominantly, instances of such pile-up (Figure 9).

The second way to extend analogy is to develop it. By complicatedly adding to Roget a set of structured knowledge-glossaries, and then using an inference-schema to pass to and fro between them, we can, again in principle, "develop the analogy in all its aspects". For instance, you can match a glossary of the parts of the body of a wolf with those of a man; provided that somebody has remembered to store the two glossaries in the machine in the first place. The actual matching algorithm, however, is not difficult to automatize; and it is this possibility of algorithmic aspect-development which makes of analogy a conceptually predictive (and therefore a scientifically predictive) vehicle (Figure 10).

Such development also makes of it a verificational vehicle; whereas analogical pile-up, on the contrary, tends to make the aspect-attributions progressively more abstract and metaphysical. For, given that the whole operation of the model has already given the analogy an underlying context by embedding it in a text, it is not unreasonable to suppose that any sentence which can be constructed, as an aspect-development of some original analogy, can be judged, in a two-valued and neo-Keynesian (Keynes 1921: Ch. 18) manner, as either contributing to the total *positive analogy*, or to the total *negative analogy*: thus providing a primitive mechanism, even at the early stage when there is no other mechanism, for confirming or disconfirming a Kuhnian paradigm.

Notes: Quite apart from the mathematical peculiarity of creating an extra lattice-point for each fan of the text (to enable the fans of the text to be mapped on to the semantic area-system of the whole *Thesaurus*), the reader should be warned that any attempt to create all possible meets and joins of a set of lattice-minimals generates an infinity, namely the "free lattice", for any set of minimals > 2. To retain finiteness, the neat thing to do is to create a lattice-sum (i.e. two lattices side by side, one with the minimals at the top and consisting of meets, and the other – as here – with the minimals at the bottom and consisting of joins), though this halves the number of inclusion-relations which can be obtained.

On the other hand, a piece of free lattice, operationally grown, never really did anybody any harm – see Figure 4. Note also, by the way, the tendency to excessive overlap, as in metaphysics – see Figure 5.

Figure 10   Analogy-development, to be achieved by turning Roget's Thesaurus *into a structure reminiscent of Bishop Wilkins'* Character Universals

(a) The reiterative pattern of the text

(b) The structure of the inference schema

(c) The lattice formed from the reiterative
     pattern, limited by the structure

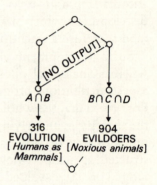

(d) The attribute-characteristic cross-over
     permitted by the schema

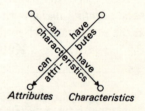

(e) The aspect-markers limiting the range
     of the schema

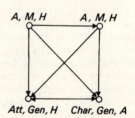

(f) Application of these to the knowledge-glossaries, to permit their enlargement, if necessary by making an aspect-search right through the *Thesaurus*.

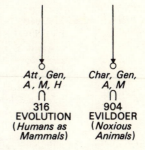

Att, Gen,
A, M, H

316
EVOLUTION
(*Humans as
Mammals*)

Char, Gen,
A, M

904
EVILDOER
(*Noxious
Animals*)

(g) [Algorithmic comparison of 'knowledge-glossaries' to form 'clicks']

The principle is to attach data-bases, or 'knowledge-glossaries' to the relevant semantic areas, and then permit selection and combination of the 'pieces of knowledge' within them by applying the rules given in the relevant inference-schemata.

This should produce a very large sequence of analogical reiterations which (with the aid of a module for generating correct English syntax) can be turned into text.

That this has not been done, although we now have machines to do it with, pinpoints the difficulties which confronted Bishop Wilkins.

Each piece of reiterative text produced by analogy development, will be called a *CLICK* (n.b. the clicks given below are illustrative only: their production has not been algorithmized).

Examples of analogy-clicking from (MAN) (IS) (A(WOLF)).

CLICK 1.1
(MEN) (HAVE (FACES))
(WOLVES) (HAVE (MASKS))

CLICK 1.2
(MEN) (HAVE (TEETH))
(WOLVES) (HAVE (FANGS))

CLICK 1.3
(WOLVES) (HAVE (TAILS))
(MAN) (HAS) (A (COCCYX))

CLICK 1.4
(WOLVES) (HAVE (FUR))
(MAN) (HAS (SKIN))

etc. etc.

CLICK 2.1
(WOLVES) (CONGREGATE)
                    (IN (PACKS))
(MEN) (CONGREGATE)
                    (IN (SOCIETIES))

CLICK 2.2
(WOLVES) (HUNT + DOWN)
                    (THEIR (PREY))
(MEN) (MAKE + WAR + ON)
                    (THEIR (ENEMIES))

CLICK 2.3
(WOLVES' (FUR)) (GOES + WHITE)
(IN (THE (WINTER)))
(MEN'S (SKIN)) (GOES + WHITE)
(IN (THE (WINTER)))

etc. etc.

Explanation of the inference process

The locations of the 'knowledge-glossaries' is found by taking the meets of the two side-chains of the lattice formed from the reiterative pattern, the arrows in this having been previously limited by reference to the schema (see (a), (b), (c) above).

IN
PRINCIPLE   These locations, 316 EVOLUTION (*Humans as Mammals*) and 904 EVILDOERS (*Noxious Animals*) would presumably have to be reached by finding common Cross-references in 371 MANKIND and 365 ANIMAL; similarly as between ANIMAL and VIOLENCE, ANIMAL and MALEVOLENCE or MALEVOLENCE and VIOLENCE; and by failing to find any other common cross-references leading to other knowledge-glossaries. (See also Fig. 4.) It is to be presumed also that the 'meet' of the diagonal chain in the lattice, $A \cap B \cap C \cap D$ will lead to no knowledge-glossary. The 'cross-over' of the schema permits *Noxious Animals* to have attributes and *Humans-as-Mammals* characteristics; and the application of the Aspect-Markers permits enlargement of the glossaries (see (d), (e), (f) above).

IN FACT   Owing to the difficulty of causing an abstract semantic idea-system, like Roget, to retrieve very concrete and detailed knowledge-glossaries, analogical clicking is in fact done by starting with exactly the right glossaries (i.e. labelled physiological diagram of a man and a labelled physiological diagram of a wolf, and algorithmically generating reiterative analogical 'CLICKS' from them.

3   Of course, developing an analogy which will be usable in science is by no means as simple as comparing a man with a wolf; and of course the ways in which analogy is used in science are by no means as easy to model as an analogy which can be developed by forming a single text down a page.

In order to gain more knowledge by example of how analogy was actually used on scientific occasions, I devised a schema representing analogy-development which, by courtesy, could have been called a two-person game. The game was between the proponents of the analogy, who kept trying to add to the positive analogy, and the opponents of the analogy, who kept trying to add to the negative analogy, in order to "shoot the whole analogy down in flames". There were three sorts of move which could be made by either side: (1) *posing* an analogy, or *composing* an analogy-cluster, the cluster being produced by constructing pile-up; (2) *clicking* an analogy, to produce aspect-development, any instance of which could then be judged to add to either the positive or the negative analogy; (3) *cannoning*, an end-game strategy performed by superimposing whole analogy-clusters, and of which further explanation is produced below. The moving-system was as in billiards, with the proponents of the analogy having the first move, and going on composing and clicking until no more positive analogy could be produced. The opponents, for their move, produced and developed negative analogy, and also produced tests "to test the strength of the analogy", which, if the tests failed, would bring the whole analogy down. There was a scoring system: every positive click scored 2; every click scored −2; and if the score became negative, the game was provisionally lost. However, it could be retrieved again, by successful cannoning. For cannoning was the bringing to bear, *on the original concrete analogy*, of a whole new analogy-cluster, with a whole new mathematical or other technique attached; and a successful cannon, which could be achieved by either player, doubled either the positive or the negative score.

I played the game on a simplified form of the kinetic theory of gases, taken from the historical chapter of Sir James Jeans' book (1921), and, in more detail, on the theory of continental drift. The composed analogies, in the first case, were those of a Democritean atom, of a set of rubber billiard balls bouncing in a box, of billiard balls ricochetting on a 2-dimensional billiard table, at first finite, then infinite; and finally, of a set of interacting points. From that stage on, the mathematics of the theory took over; but it was interesting to see, in Van der Waal's correction, the original concrete analogy had to be reasserted, and the points reconverted, in imagination, back into balls. But the final cannon, which established the theory and won the game, only occurred quite late on, with the discovery of Brownian movement. Until then, there were far more opponents of the theory, producing negative analogy, than we now realize.

In the second case, that of continental drift, and where there were no mathematics, far more weight was thrown on to the actual game. Here the analysis was taken from a *New Scientist* paper by Barrie Jones (1974), called 'Plate Tectonics, a Kuhnian Case?' The analogies used by the proponents of the theory were the jig-saw puzzle analogy, with aspect-development showing continuity of animal-plant pattern: the blown-up balloon analogy, which made the evolutionary movement of the earth's crust able to blow up vertically, with wrinkles for mountain-ranges: and a toy-boat analogy (disastrous, but implicit in the very word 'drift'), which, on a stabilist earth, allowed the continents to float about.

The opponents of the theory demolished the toy-boat analogy, by causing it to produce any number of negative clicks, and they produced yet more by adducing instances of "misfit" of the jig-saw puzzle. In addition, they produced, to replace the toy-boat analogy, an intercontinental bridge-formation analogy which they knew would prove untrue, in that no traces would ever be found of any of the bridges. They further produced negative clicks from the blown-up balloon analogy. All this caused the score to become negative, and so the theory was judged to have been shot down in flames.

However, much later, plate-tectonics re-established the theory by producing a famous and successful cannon. The new analogy-clusters, then brought to bear were: the acoustic analogy of shape-determination by echo (with technique attached); the analogy of handlike mechanisms moving to juxtapose symmetric/asymmetric remanent magnetic patterns; the analogy of the unevenly rising pie-crust (which replaced that of the expanding toy balloon); the analogy of the twisting plasticine, the twist being caused by the wandering of the earth's magnetic poles; and the analogy of over-lapping iron plates, which piled up with an analogy of similarly overlapping large flakes on the pie-crust to produce a picture of large-scale horizontal movements on the earth's crust. All this cannoned on to, and saved, the jig-saw puzzle analogy; because it created, by aspect-development, such a new wealth of positive analogy that it was judged that the negative analogy (all of which still remains), would never catch up, and so the game was judged won.

This kind of analogy-handling, even though only carried on in two cases and in a simplified way, convinced me once and for all of the absolutely cardinal role which analogy-handling (once you have the model with which to handle the analogies) can be shown to have in all paradigm-based science; and since then, examination of other, more sophisticated cases, such as that of black holes in the universe (see Clarke 1978), has strengthened this belief yet more.

The handling also prompted the reflection that analogy is (over-all) used in converse ways, when it is used in poetry and when it is used in science. For in poetry, we tend to apply novel, hitherto undreamed-of analogies to illuminate ordinary, easily perceived or intuited circumstances, so that we add to our ordinary vision a new vision, which makes us see ordinary circumstances in a new way. In science, on the contrary, we pile up quite ordinary, not to say obvious analogies, and then apply them counter-intuitively, in extraordinary circumstances, so that the analogy itself supplies our only guide to further exploration in the conceptual darkness into which we have then projected ourselves. Thus, as an example of the first, consider Cecil Day Lewis' poets' analogy (though actually to be found in a prose work): 'The barrage balloons, those whales, floated over London.' Here the analogy itself is not ordinary; for whales are esoteric, not normally observed creatures. And yet how good it is. For having read the sentence, together with its metaphor, almost any ordinary person who has ever observed barrage balloons, will now be able to "see" also the great aerial fishes, floating and swaying over the besieged town. Contrast this with the use of analogy in the kinetic theory of gases. This employs quite an ordinary pile-up of analogies, of rubber balls bouncing, or billiard balls ricochetting. Yet, until Brownian movement became observable through microscopes, (which is actually to posit yet another analogy, since, with an ordinary microscope, the actual molecules are much smaller than the entities which we appear to observe), who, using their unaided common sense, would ever have imagined that gases were like that? And since no ordinary person could have produced the act of imagination, no ordinary person, either, could have invented kinetic theory.

Nevertheless, in the theory, the pile-up was there; and thus the fact has got to be faced that Kuhn, by digging for paradigms, saw, in effect, the absolutely cardinal fact that, in real science, the thinking which goes on is enormously cruder, with the crudity going on much longer, than anyone who had not actually observed it going on would have believed possible. And why? Because the whole purpose of inventing scientific theory is to get some strongminded way – any way – of predicting what is going to happen in totally unknown areas of reality which you cannot in any other way explore; and of then verifying to see whether it then does happen or whether you have got to start again. And that is what Braithwaite saw (1953: 263f), and Kuhn, in effect, did not.

Cambridge Language Research Unit

REFERENCES

Black, Max. 1962. *Models and Metaphors*. Ithaca.
Braithwaite, R. B. 1953. *Scientific Explanation*. Cambridge.
Campbell, N. R. 1920. *Physics the Elements*. Cambridge. Later reprinted as *Foundations of Science*. New York.

Clarke, Chris. 1978. Black holes. *Theoria to Theory* **12**, 275–9.

Hacking, Ian. 1975. *Why Does Language Matter to Philosophy?* Cambridge.

Hesse, Mary. 1974. *The Structure of Scientific Inference*. London.

Isham, Chris. 1979. Quantum gravity. *Theoria to Theory* **13**, 19–27.

Jeans, J. H. 1921. *The Dynamical Theory of Gases*. Cambridge.

Jones, Barrie. 1974. Plate tectonics: a Kuhnian case? *New Scientist* **63**, 536–8.

Keynes, J. M. 1921. *Treatise on Probability*. Cambridge.

Kuhn, T. S. 1962. *The Structure of Scientific Revolutions*. Chicago.

Masterman, Margaret. 1961. Translation. *Aristotelian Society Supplementary Volume* **35**, 169–216.

Masterman, Margaret. 1970. The nature of a paradigm. In *Criticism and the Growth of Knowledge*, ed. I. Lakatos and A. Musgrave, pp. 59–89. Cambridge.

Roget's Thesaurus. 1962 edition. London.

Wilkins, John. 1668. *An Essay Towards a Real Character and a Philosophical Language*. London.

Wilks, Yorick. 1975. A preferential pattern-seeking semantics for natural language. *Artificial Intelligence* **6**, 53–74.

5 A revised regularity view of scientific laws

MARY HESSE

1 What is the relation between regularities, successful factual prediction, counterfactual inference, and lawlikeness? If one accepts that scientific theory not only licences but requires successful factual prediction and counterfactual inference, is one also obliged to accept that scientific laws are modal statements carrying some kind of necessity? It is received wisdom that successful factual prediction and counterfactual inference are required, and so far I share received wisdom (Hesse, 1962), but I deny that it follows that laws are statements of necessity. I shall therefore explore a revised version of the regularity view drawn from the tradition of Hume, Mill, Carnap, and the so-called Cambridge school of Johnson, Keynes, Ramsey, Broad, Jeffreys and Braithwaite (Pietarinen, 1972).

In a recent reconsideration of "Laws of Nature", Fred Dretske (1977) has conveniently summarized the required features of laws as follows:

(a) A statement of law has its descriptive terms occurring in opaque positions, in the sense that replacement of a descriptive term by another which happens to be coextensive with it does not in general preserve the lawlikeness of the original although it does preserve its truth.

(b) That a statement is a law is an objective matter independent of epistemic considerations.

(c) (i) Laws can be confirmed by their instances.
(ii) Confirmation of a law raises the probability that unexamined instances will resemble (in the respect described by the law) the examined instances. Hence laws have predictive force.

(d) Laws figure in the explanation of phenomena within their scope.

87

(e) Laws "support" counterfactuals.

(f) Laws tell us what must happen, not merely what has happened and will happen.

Dretske characterizes a standard view of laws as the thesis that laws are "universal truths plus X", where X indicates the extra features such as (c), (d) and (e) that a universal generalization must have if it is to function as required in scientific systems. This he calls the *functionalist* view. He argues that the candidates (c) to (e) for X are all epistemic, that is depend on our knowledge, whereas laws should be objective. Moreover he concludes that it cannot be shown how any X could endow any universal generalization with either predictive, explanatory, or counterfactual force. Neither, he thinks, does it help to insert a modal connective representing necessity in the formulation of the law, since the nature and justification of such a connective remains entirely unexplained however rigorously its modal logic may be fashioned to satisfy conditions (a) to (f). This necessitarian version of the standard view 'is not a solution; it is part of the problem' (Dretske 1978).

Dretske proposes to replace the standard formulation

$$\text{law} = (x) \ (Fx \supset Gx) \text{ plus } X,$$

where the law is construed as an intensional relation between extensions, by an *extensional* relation between *intensions*, that is between properties or universals. The formulation becomes

$$F\text{-ness} \rightarrow G\text{-ness}$$

This, in Dretske's terminology, represents an "ontological ascent" from talk about instances and factual generalizations to talk about universals. Dretske says he attaches no special significance to the connective '$\rightarrow$', since it will depend on the type of law concerned, that is whether it is quantitative or qualitative *etc.* But it is clear from his insistence that the relation is *extensional* that '$\rightarrow$' is not supposed to indicate any modality, but merely extensional copresence and covariation.[1] And yet because of the universal character of F-ness and G-ness, it follows that *particular Fs must be Gs* in virtue of exemplifying the universals. Laws as relations between universals "*go beyond* the sets of things in *this* world" that exemplify these universals, hence laws can be seen to satisfy conditions (a) to (f) (Dretske 1977: 266).

---

[1] Niiniluoto (1978: 436–7) seems to be mistaken when he suggests that Dretske should further define the significance of his connective '$\rightarrow$'. It is not this connective that is problematic in Dretske's account, but rather the universals connected by it.

The trouble with Dretske's suggestion is of course, as Illka Niiniluoto points out in his reply, that he does not explain what he means by universals nor how they can be connected extensionally. Niiniluoto responds with a restatement of the view of laws as necessary connections, but this response must be subject to the very same type of objection, namely that the nature (as opposed to the functions) of this connection is not explained. The functions remain to be justified; they cannot themselves act as explanations of the nature of necessity. Indeed there is a close parallel between the kinds of defences and objections both authors put up for their respective pieces of metaphysics: Dretske appeals to universals that go beyond what is asserted by universal truths "of *this* world", Niiniluoto appeals to the "class of physically possible worlds". Dretske speaks of "ontological ascent", Niiniluoto of "nomic necessity". But all these concepts are described merely as those which *are capable of* satisfying conditions (a) to (f) by fiat. It follows that they cannot throw more light on these conditions, much less explicate them as belonging to a non-circularly justified coherent system. Both views are in fact versions of the same view, namely that laws give us access to *more than this world. How* we know of the existence of this "more" and how we gain access to it remain equally mysterious in both views.

2   Let us go back to Dretske's conditions (a) to (f) and try a version of the regularity view again. I shall accept the need for laws to satisfy all these conditions except the last, for this is merely a statement of some view that entails that the regularity view is false, and exactly what view is imported by the word 'must' is not clear until some non-regularity view is spelled out. I am not going to defend the regularity view in the functional form in which Dretske discusses it, namely as 'law = universal truth plus X'. I agree with both Dretske and Niiniluoto that this view is inadequate, though not for all the same reasons.

Consider first condition (b). This is unexceptionable, but it is doubtful if any proponent of the regularity view would agree that he violates it. If a law is a universal generalization which can be confirmed, related deductively in a theoretical explanatory system, and is successfully predictive, then on the regularity view these empirical and logical properties of the generalization are facts, and are just as objective and independent of our knowledge as any other facts. That a generalization has these properties *makes* it a law on the functional view, but these properties are not themselves merely epistemic, although they depend for their acceptance *as* properties upon epistemology. So does the fact that Jupiter has moons, and indeed what acceptance as true properties of an object does not? Perhaps, however, what Dretske is objecting to in proposing condition (b) is the notion that by merely stating epistemic requirements of certain sorts one has shown how to turn a factual truth into something necessary or metaphysically universal. But regularity theorists do not (generally) claim to do this; they are satisfied with laws as facts with certain properties and functions.

Condition (c) is more difficult. I shall distinguish two parts of this condition: (i) referring to confirmation of laws, and (ii) to confirmation of unexamined instances. I have argued elsewhere that (i) is irrelevant to the predictive power of science, and that (ii), in the form stated by Dretske, commits the transitivity fallacy.[2] This fallacy arises because there is no non-vacuous confirmation theory in which one can argue transitively as follows: positive instances confirm their generalization; therefore the positive instances confirm a prediction that the unexamined instances will satisfy the generalization. In particular this is not true on either of the two standard construals of the relation 'confirms', that is the so-called positive relevance criterion or the k-criterion. To satisfy the very reasonable demand that something *like* this transitivity condition be true, further constraints are required. Among various suggestions as to what these further constraints should be, two are particularly relevant to the problem of lawlikeness.[3]

The first of these is developed by Niiniluoto and Tuomela (1973). (Let us call them and their theory 'N–T'.) They replace the transitivity of confirmation from observed instances to predictions by what they call "hypothetico-inductive" inference from observed instances *plus accepted background theory T* to predicted particulars. In

---

[2] *Cf.* my (1974, hereafter "*SSI*") pp. 141–6. Although I argued in Chapter 8 of that book that it is reasonable to suppose that universal generalizations in an infinite domain have probability zero, nothing in the present paper will depend on the correctness or otherwise of that argument. All that is needed here is the weaker assertion that the confirmation of universal generalizations is not *necessary* for the confirmation of their predicted instances.

[3] There is another important suggestion, namely that if the confirmation of a generalization by instances can be shown to increase monotonically to 1, then the transitivity condition is satisfied to as good an approximation as required by increasing the number of instances examined. But it is still a controversial question whether the theorems required to show this are applicable to typical cases of inference in science. See Dorling (1975: 61; and 1976: 160).

effect, T is accepted with maximal confirmation value for purposes of an inductive inference from observed to predicted particulars. Theories are said to establish inductive systematization in so far as they are *required* to induce particulars from particulars, and the measure of systematicity is the measure of their lawlikeness. But how do theories come to be part of the accepted background premisses? N–T reply

It is not necessary . . . to take any definite standpoint in answering the question of the origins of the hypothetical theories occurring as premisses of hypothetico-inductive inference. They may be conjectures which have more desirable features than their rivals – for example, higher content, greater explanatory power, or higher degree of corroboration on the basis of higher-level theories. They may be independently testable, deductively or inductively. They might also be theoretical presuppositions only, which are considered as unproblematic and tentatively accepted in some situation (1973: 210).

These seem to be feeble and circular grounds on which to rely for predictions upon which action is to be taken, especially in circumstances in which the negative utility of failure is very high (*cf. SSI*: 198 and 207). And in the present context of the analysis of lawlikeness, N–T have not given anything other than a "functional" account of laws and theories. There is the same circularity of lawlikeness, confirmation, acceptability, and explanatory power that Dretske points out in relation to the "universal truths plus X" construals of laws. Because of this circularity, N–T not only fail to give an adequate account of lawlikeness, but also fail to satisfy even the requirement of confirmation of predictions which is the core of Dretske's condition (c). (*Cf.* my (1979), and Niiniluoto's reply and the subsequent discussion).

The second method of avoiding the transitivity fallacy to be considered here is the one I developed in *The Structure of Scientific Inference*, namely the suggestion that what ensures confirmation of particulars by appropriate particulars is not confirmation *via* universal generalizations (or laws, if such there be), but is a relation of similarity between the particulars. Generalizations that are candidates for lawlikeness on this view become statements in a finite domain to the effect that certain examined particulars exhibit pairwise similarities in certain respects that licence prediction. Dretske's condition (c) therefore has to be replaced by:

Examined instances of the generalization 'All Fs are Gs' raise the probability that unexamined Fs will resemble (in being Gs) the examined instances.

This reformulation is still not quite right, however, because it neglects a crucial *ceteris paribus* clause. Given, for example, that all trees in temperate regions shed their leaves in winter this evidence should not necessarily raise the probability that all trees in tropical regions shed their leaves in winter. The condition should therefore be reformulated as:

(c') Examined instances of the generalization 'All Fs are Gs' raise the probability that unexamined Fs that are sufficiently similar in other relevant respects not known to include G, will be similar to the examined instances also in being G.

The inference is an analogical inference from instances to instances, and rests on a general postulate to the effect that the universe is to some degree homogeneous (the "clustering postulate"). This postulate constitutes a constraint on the initial probability distribution presupposed by the confirmation theory, thus ensuring that (c') is satisfied (*SSI* Ch. 7).

The crucial difference between this view and the N–T account can best be indicated by noting that N–T are forced to give quite distinct accounts of theories as legitimations of predictions on the one hand, and predictions as conclusions of analogy arguments from particulars on the other. Both are legitimate forms of inference in their theory, but far from operating in terms of similarities or analogies, theories have *no* systematizing or predictive force for N–T just in those cases where there are analogies and hence inductive inference between particulars. This means, for example, that if the theory of gravitation is to have inductive systematicity according to N–T, there must be *no* confirmation of the inverse square force for comets and planets, because there is only systematization *in the absence of* confirmation by analogy between particulars (Niiniluoto and Tuomela 1973: 10).

In order to justify the similarity account, however, we have to elucidate the reference in (c') to 'sufficiently similar in other relevant respects'. It is often argued by necessitarians that this cannot be done in a regularity view without importing some assumptions about what sorts of predicates F and G can figure in laws, and what are mere accidental correlations. For example, we do not know how similar in what respects to a given sample of pure water a liquid has to be to suit a car battery unless we know relevant chemical laws, and chemical laws must imply more than a set of other generalizations of the kind 'All *F*s are *G*s', for which on the regularity view the same problem of sufficient similarity of instances would arise. Or again, it was once noticed that the sizes of the orbits of all known planets satisfied a simple algebraic formula called 'Bode's law', but it did not follow that any unexamined planet, very similar in all other respects, would have a higher than initial probability of satisfying that formula, since there was no reason to suppose the formula to be law.

It is not clear, however, that such objections are conclusive against the regularity view. Judgements of relevance and irrelevance can themselves be grounded on further tests for extensional correlation and regularity. Consider the regularity 'All free massive bodies dropped from rest in the neighbourhood of the earth fall with approximately constant acceleration.' This was known long before the explanatory theory of universal gravitation was established. It was known that the conditions 'free', 'dropped from rest', and 'in the neighbourhood of the earth' were *relevant* conditions (the moon, comets, arrows, smoke, birds, do not satisfy the regularity). It was also known that the size, shape and material constitution of the bodies, and their latitude and longitude as such, were *irrelevant* conditions (Galileo, Torricelli, Pascal and their successors were concerned to check the details of these conditions). The arguments for irrelevance were of the same kind as those for relevance, namely experiments exhibiting agreement and difference and covariation such as were laid out in the classic inductive methodology of Bacon, Hume, Herschel and Mill. 'Sufficient similarity in relevant respects' can be cashed without circularity in terms of other known regularities.

The example of Bode's law can also be dealt with in terms of regularities. Why do we not suppose it to be a law? Because it asserts a regularity in the spatial positions of particular bodies, and this is not the type of relation actually found in connection with the mechanical behaviour of other material bodies. The domain of instances "sufficiently similar in other relevant respects" has to be extended beyond planets to all massive bodies, and in this case evidence of seven planets is not sufficient to outweigh the *negative* evidence from all other bodies that there are no regularities in their mere positions in space.

But this kind of reply does not solve the more fundamental difficulty of specifying the nature and degrees of similarity that are presupposed by it. The wider the range of phenomena that have to be taken account of in the evidence, and the more "holistic" the initial probability distribution therefore becomes, the more difficult it becomes to define degrees of homogeneity of the universe. Indeed even in the simple first-order predicate languages considered by Carnap and Hintikka, specification of a probability distribution increasing with overall homogeneity of the universe turns out to involve intractable combinatorial problems, especially if relational predicates are introduced. No reasonably finite processing mechanism, brain or other, would be large enough to contain all the possibly relevant information. What Suppes has called the "combinatorial jungle" must somehow be tamed (Carnap 1950: 124; and Suppes 1966: 41).

There is further and more fundamental objection to the notion of similarity, to the effect that subjective judgements of similarity are themselves dependent on theories, and that, as an objective relation, similarity is only strictly definable in terms of a true theory of the world. Thus Quine says:

A man's judgments of similarity do and should depend on his theory, on his beliefs; but similarity itself, what the man's judgments purport to be judgments of, purports to be an objective relation in the world. It belongs to the subject matter not of our theory of theorizing about the world, but of our theory of the world itself. Such would be the acceptable and reputable sort of similarity concept, if it could be defined . . . Chemistry . . . is one branch that has reached this stage. Comparative similarity of the sort that matters for chemistry can be stated outright in chemical terms, that is in terms of chemical composition. Molecules will be said to *match* if they contain atoms of the same elements in the topological combinations . . . In general we can take it as a very special mark of the maturity of a branch of science that it no longer needs an irreducible notion of similarity and kind (Quine 1969: 135, 138. *Cf.* Goodman 1970: 19 and my reply in *SSI*: 66–70).

If this were entirely correct, it would render the present similarity account of predictive inference circular. For we should have to accept the judgements of similarity imposed by our theories, while the grounds for accepting the theories, and the predictions entailed by them, consist of these same judgements. There is one immediate consideration that shows that Quine's account cannot be *entirely* correct, and that is the impossibility of knowing in any single case of, say a chemical substance, *what substance it is*, and therefore what molecules it is composed of according to the theory, unless there are *some* primitive judgements of similarity ('There's gold again . . .') that are independent of developed systematic theory. Again, as Quine states his analysis of natural kinds, it does seem to depend on an unnecessarily strong assumption of the "convergence" through time of fundamental scientific theory, and upon the notions of an ultimately true theory and ultimately correct set of natural kinds. Recent discussion of the revolutionary character of theory-change, particularly in respect of reductionist theories like physics and chemistry, has shown this assumption to be highly problematic.

It remains true, however, that an adequate confirmation theory has to do justice to the self-corrective character of the relation between theories and inductive inference. On the other hand low-level argument from similarities is often an approximate, messy, and highly inconclusive preliminary to the establishment of theory in precise and if possible simple mathematical terms. Perceived similarities and the respects in which similarity is found to be significant are often modified and corrected by the development of theory. On the other hand, the process does depend on there being *some* residual perceptual similarities which serve as checks on the natural kinds that are eventually accepted as the best correlations of the laws of a given domain. Thus, flower colour is an obvious perceptual similarity, which may define suitable classes of plants for certain purposes: camouflage, floral display, *etc.* But if the aim is the scientific one of discovering a structure of comprehensive laws connecting plant forms with genetics, biochemistry, and the theory of evolution by natural selection, then flower colour becomes a comparatively unimportant feature in defining relevant natural kinds. It gives way to other similarities of stamen-structure *etc.*, which were not superficially obvious. The taxonomic theory still depends on some observable similarities, even if they would not have been noticed without the influence of the developing theory. The process is not circular but internally self-corrective.

Since judgements of relevant similarity should be subject to correction by theory, we should not expect a definition of fixed kinds and degrees of similarity to be built into the initial probability distribution for all time. Rather the lessons of past theory should be reflected in a continually modified initial distribution. The question is how to do this without either accepting past theory as incorrigibly true, or falling into a transitivity fallacy. I argued above that Niiniluoto and Tuomela's method of hypothetico–inductive inference is not satisfactory as it stands, but we may take a hint from their use of "background theory". This notion has been much discussed by Popper and his successors, but the actual content of the background is not often examined in detail. Does it consist of classical theories successful in the past, and if so which? Surely not such theories as Newton's or Maxwell's or Dalton's, at least not without heavy qualification, since these theories have been found to be strictly false. Does it consist of lower-level generalizations which more or less persist through changes of higher-level theory, such as Boyle's law, Fresnel's laws, the laws of constitution of particular chemical compounds? Perhaps, but these would also have to be hedged about with statements of approximation, and would also raise problems about terminology, since the way in which they are stated frequently depends on the theory-laden connotations of their constituent terms, and therefore changes from theory to theory. It is difficult to see how general background theories can be specified so as to be both acceptable as true, and strong enough to perform the function required in the N–T theory.

We have, however, already assumed in our reformulated condition (c′) that we are concerned principally with inference from instance to instance, and hence by implication with inference from and to finite conjunctions of instances. It follows that, as far as (c′) is concerned, we can take 'All Fs are Gs' in the weak form of a finite conjunction of observed and unobserved instances. The statement of the generalization is not required to be of infinite scope; its significance is rather that it can be taken as lawlike in so far as it is both a true statement of regularity, and carries the presupposition that what count as Fs and Gs are respectively clusters of relevantly similar instances. Now let us apply the same treatment to "background theories". What is objectionable about the idea of *accepting* these with maximum probability value is that they are usually assumed to be universally quantified over past and future times, and hence to have *specific future reference*. This is too strong an assumption for a confirmation theory that is precisely concerned with *probability* distributions over a problematic future. But if we treat statements of background theory in the same way that we have

treated the statements of regularity 'All *F*s are *G*s', that is as summaries of past evidence of *F*s and *G*s together with a probabilistic
projection depending on perceived similarities to future instances,
this objection is removed. We may then short-circuit the problems
of formulating a general clustering postulate over total past evidence by adopting into the evidence background theories only as
convenient summaries of what have been found to be successful
"natural kind" classifications in the past. They then have no *entail-
ments* for the future, but they do provide a structure of assumptions
for the conditional probability distribution over possible future
states. The only similarities involved are perceived similarities,
since we are concerned with applications to observable future
instances, but they are similarities of kinds that have been modified
and selected out of the total evidence by theoretical processing.
What have turned out to be irrelevant similarities have been dis-
carded from the total evidence.

On this understanding a detente can perhaps be reached between
the N–T theory and the account of instance confirmation by
similarities. It may indeed not be an unwelcome suggestion for
N–T, since it shows how to combine two considerations that
remain separate in their account, namely the specification of their λ
and α parameters indicating degrees of homogeneity of the universe
in the initial probability distribution on the one hand, and the
background theories accepted as part of the evidence upon which
past probabilities are conditionalized on the other. Treated this
way, background theory summarizes whatever it is that Quine
believes is carried forward from past science when he suggests that
we have learned better and better specifications of "natural kinds"
from that science. But it does not carry the implication apparently
accepted by Quine that the structure of natural kinds is now relatively fixed in advanced sciences. Judgements of similarity between
past and future instances may continue to take weak perceptual
form, so long as background theory is allowed to educate our
judgements of past similarities by processing the evidence upon
which probabilities of future instances are conditionalized.

This account needs to be worked out in more detail and in
application to particular examples. But technical difficulties apart it
is enough to show that condition (c') can be satisfied without resort
to assumptions about necessity or essentialist natural kinds. If the
probabilistic basis of this account is accepted, the modal element

said to be required in an account of scientific laws can be understood as probabilistic rather than necessitarian. It has the advantage over modal necessity that it forms part of a well-accepted empiricist analysis of actual and possible *facts*. Talk of physical necessity and physically possible worlds is replaced by talk of physical *contingency*, *logically* possible worlds (the worlds described in the basic predicate language), and a *probability* distribution over these worlds. What is discovered in the development of theory is not necessities, but contingent facts about the comprehensive character of various kinds of simple regularities and similarities among phenomena. It is no doubt a fortunate fact for the progress of science that comparatively simple correlations have been found to be as widespread in the physical sciences as they are. Since this is a contingent fact, however, there is of course no guarantee that such degrees of clustering will be repeated indefinitely either in physics or the other sciences. We have already gone far beyond what seems to be the degree of simplicity of the world required for our own biological evolution, and it may be that we are approaching the limits of contingent simplicity required for further scientific advance.

3 We have now to consider whether this revised regularity account satisfies Dretske's other conditions for lawlikeness. Take, first, condition (a), of *opaqueness*. Descriptive terms should occur in opaque positions with regard to the lawlikeness of a statement of regularity. Suppose F is 'is a diamond', G is 'has a refractive index 2.419', K is 'mined in kimberlite' and F and K are coextensive. Why does '"All Fs are Gs" is a law' not entail '"All Ks are Gs" is a law' (Dretske 1977: 250)? Elucidation of this question by means of the revised regularity account will provide at the same time an example of how inference from similarity works in general.

The *prima facie* reasons for opaqueness are based on the following considerations: the class of crystallized minerals is characterized by having a refractive index specific to each mineral, but it is not the case that each mineral of the class is found in coextensive relation to any particular kind of rock, nor are all minerals mined in particular sorts of rock found to have the same refractive index. These facts constitute part of our total evidence. In order to translate this argument into the regularity account we must first interpret the "lawlike" character of the F–G relation as meaning that the probability of the next F being G is raised by the evidence, whereas the probability of the next K being G is either not raised or not raised to the same degree. Let us suppose that what successful-in-the-past theories tell us is that observed Ms have always been Ns, where M is 'is an instance of the class of crystallized minerals' (including instances of diamonds, opals, sapphires, emeralds. . .), and N is 'has a definite refractive index specific to their type of crystallized mineral'. There are strong pairwise relations of similarity between all the Ms. There is therefore high confirmation of the prediction 'The next M will be N', and in the absence of any evidence that Fs are untypical Ms, also of the prediction 'The next F will be N'. Since in the evidence all Fs have been Gs, it follows that there is high confirmation of 'The next F will be G'.

All such informal probability arguments must incur at least the suspicion of falling into some form of transitivity fallacy. It is therefore necessary to go through the last step of the above argument in some detail to clear it from this suspicion. Suppose our successful-in-the-past theories have told us that the relevant part of the total evidence is

$E_1 \equiv$ 'All Fs have been Gs'
$E_2 \equiv$ 'All Ms have been Ns'

If x_i is the next instance, we have

Fx_i & $Nx_i \equiv$ 'x_i is F and has the same value of refractive index as other Fs'
Fx_i & $Gx_i \equiv$ 'x_i is F and has the value, say, g of refractive index'.

Then we have

Fx_i & Nx_i & $E_1 \equiv Fx_i$ & Gx_i & E_1.

We assume Nx_i is confirmed to a greater degree by Fx_i & E_1 & E_2 than by Fx_i & E_1 alone, that is

$p(Nx_i / Fx_i$ & E_1 & $E_2) > p(Nx_i / Fx_i$ & $E_1)$

But $p(Nx_i$ & Fx_i & E_1 & $E_2) = p(Gx_i$ & Fx_i & E_1 & $E_2)$
and $p(Nx_i$ & Fx_i & $E_1)$ $\qquad = p(Gx_i$ & Fx_i & $E_1)$.

Therefore $p(Gx_i / Fx_i$ & E_1 & $E_2) > p(Gx_i / Fx_i$ & $E_1)$.

That is, Gx_i is confirmed to a greater degree by Fx_i & E_1 & E_2 than by Fx_i & E_1 alone.

On the other hand, in the case of the prediction 'The next K will be G', we have no wider class of instances similar to K which support the prediction. That is, there is no evidence analogous to E_2, because it is not the case that all minerals found in particular sorts of rocks have the same refractive index.

Thus, to cut a long story short, opacity is ensured in the revised regularity account by the fact that coextension with a given class does not entail coextension with a wider class of instances similar to those of the given class. It should be noticed in particular that we have not introduced any higher-order predicate language here, not any concept of "similarity of classes". The class of Ms is not the class of classes of crystallized minerals, but the class of *instances* of crystallized minerals. Therefore the similarities involved are still pairwise similarities between individuals, for example between individual diamonds and sapphires and emeralds, and these similarities are assumed to rest on primitive observable relations processed for relevance by past successful theories and generalizations.

Dretske's condition (d) asserts that laws have *explanatory force*. On the regularity view, explanation is understood as systematization of regularities in a theory of comprehensive scope. In particular, since the account given here places the emphasis on relations between particulars rather than deduction from general theoretical premises, theory is to be viewed as a comprehensive summary of the kinds of regularity and similarity between particulars that have been found to be successful in the past. Generalizations that have been found to have lawlike properties in the past get to be included in the relevant evidence upon which predictions are conditionalized, and, conversely, successful theory in the form of that relevant evidence determines what generalizations have predictive force. In this sense the example considered in connection with the opaqueness condition shows how lawlikeness and explanatory force are related.[4]

The most crucial test of a theory of lawlikeness is how it deals with the problem of *counterfactuals*. First it should be noted that the kind of *evidence* for lawlikeness we have already discussed is generally taken to be evidence also for counterfactual force (Pietarinen 1972: 119f). That is to say, evidence for 'The next F will be a G' is taken to be evidence for

(CF)   'If an unobserved x_i were to be an F (though it isn't), then it would be a G'.

---

[4] I do not discuss here the view that explanatory theories should have reference to real but unobservable entities and processes, but see my (1977). The present account does however rest on the assumption that while there may be unobservable individuals (whether entities or events), no unobservable *predicates* can have any place in predictive inference, since we should not know how to make perceptual similarity judgements between individuals characterized by unobservable predicates. See *SSI* Chaps. 2 and 9.

Discussion about whether so-called counterfactuals actually *imply* the falsity of the antecedent is beside the point here. What is required for the analysis of *laws* is that there should be counterfactual force whether the antecedent is false or not. For example, theories are generally assumed to make lawlike assertions about ideal models which can never be actualized without violating other laws of the system (*e.g.* ideal gases, the absolute zero of temperature), and there are also what I have called 'avoidance situations' in Hesse (1962), where the law itself predicts an undesirable consequence if the antecedent is fulfilled, so that we take steps to see that it never is or will be (*e.g.* Braithwaite 1953: 305 refers to W. E. Johnson's "brakeless train").

The question is, how can evidence for 'The next *F* will be a *G*' be evidence for (CF), given that the factual prediction does not entail it? (Incidentally, even if it did entail it, that would not help, because any confirmation argument based on such an entailment would fall into a transitivity fallacy.) Here the probabilistic character of predictions in the revised regularity view comes to our aid. In a probabilistic confirmation theory a conditional probability distribution is defined over states of the universe that are logically possible given the total evidence, provided of course that this evidence is self-consistent. There is therefore a conditional probability of the form:

$$p(Gx_i \,/\, Fx_i \,\&\, E)$$

where E is the rest of the total evidence, including the instances already observed of *F*s that are *G*s, if any. Now an ambiguity appears in the formulation of (CF). It may be taken to imply 'x_i is not an *F* and never will be'. If this were added to E it would render the total evidence inconsistent and the probability expression vacuous. But it is not easy to envisage any situation in which prediction and action are contemplated that could require us to find a probability conditional upon inconsistent evidence. In particular, ideal models of the kind referred to above are not used for prediction unless accompanied by applicable conditions under which they are approximately actualized. Counterfactual inferences do affect action and prediction only in avoidance situations, that is, where x_i is not an *F* *now*, but could be made an *F* by intervention, or may naturally become an *F*, *in the future*, and where we may wish to guard against this happening. Inserting temporal parameters in the evidence will easily yield a self-consistent condition for this case. The conditional probability on which we act is the probability of

Gx_i given the temporally qualified condition 'x_i is not F now at t_0 and is F later at t_1'. If desired we may then act to ensure that 'F at t_1' is false.

There are, it is true, counterfactual inferences in science of the form 'If x_1 had been F at time t_0 (and it wasn't) then it would have been G at time t_0'. For example 'If the earth's orbit had been larger, the year would have been longer'. Such inferences are represented in the probability distribution by the conditional probability of Gx_i and given Fx_i and such of the total evidence as is consistent with Fx_i. That is to say, we contemplate a (logically) possible world at t_0 as near to the actual world at t_0 as is consistent with the replacement of the known 'x_i was not F at t_0 by Fx_i at t_0'. Again talk of physically possible worlds in the necessitarian view is replaced by talk of conditional probabilities over logically possible worlds on the regularity view. A probability value is licensed by this counterfactual conditional probability expression, but no action involving intervention or prediction is licensed or required. Therefore the requirement of total (actual) evidence can be waived in this sort of case, which is of the nature of hypothetical model-building or thought experiment, and involves only the internal consistency of a theory, not directly the truth or probability value of empirical statements.

In conclusion, it may be asked whether the revised regularity view depends crucially on acceptance of a probabilistic confirmation theory? It is possible that some other form of regularity view would do the job, but none seems to be on offer at present. In particular, the various forms of functionalism described by Dretske are inadequate because they do not exhibit any intrinsic connection between his requirements (a) to (e) on laws. A conditional probability theory is able to construe all these as features of a suitable initial probability distribution together with evidence that has been selected and processed by successful-in-the-past theories. There is indeed one non-probabilistic confirmation theory that is in some respects an accurate explication of scientific practice, that is the theory of L. J. Cohen. But this depends at crucial points upon judgements of "causal necessity" which have to be made *prior* to application of the theory, and it must therefore be considered as a species of necessitarianism rather than of the regularity view (Cohen 1970: 23f). Again the distinction between the regularity and necessitarian views is seen to be mirrored in their use of probabilities or of causal modalities.

In a revealing example, Dretske compared his pattern of inference from universal to universal, with its conclusion 'This *must* be a G', with certain features of the American Constitution. It is strange that metaphysical thinking about theories of nature from the seventeenth to the end of the twentieth centuries should come back full circle to the metaphor of the Divine Lawgiver. I hope I have shown that while He may indeed be required to dispose the *facts* in a relatively economical way relative to human intelligence, He is not required to underpin the facts with modal necessities which even He might find incomprehensible.

University of Cambridge

REFERENCES

Braithwaite, R. B. 1953. *Scientific Explanation*. Cambridge.
Carnap, R. 1950. *Logical Foundations of Probability*. London.
Cohen, L. J. 1970. *The Implications of Induction*. London.
Dorling, J. 1975. The structure of scientific inference, *British Journal for the Philosophy of Science* **26**, 61.
Dorling, J. 1976. The applicability of Bayesian convergence-of-opinion theorems, *British Journal for the Philosophy of Science*, **27**, 160.
Dretske, F. I. 1977. Laws of nature, *Philosophy of Science*, **44**, 248.
Dretske, F. I. 1978. Discussion: reply to Niiniluoto, *Philosophy of Science*, **45**, 444.
Goodman, N. 1970. Seven strictures on similarity. In *Experience and Theory*, ed. L. Foster and J. W. Swanson, p. 19. London.
Hesse, Mary. 1962. Counterfactual conditionals, *Aristotelian Society Supplementary Volume*, **36**, 201.
Hesse, Mary. 1974. *The Structure of Scientific Inference*. London.
Hesse, Mary. 1977. Truth and the growth of scientific knowledge. In *PSA 1976*, Vol. 2, ed. F. Suppe and P. D. Asquith, p. 261. Philosophy of Science Association, East Lansing, Michigan.
Hesse, Mary. 1979. What is the best way to assess evidential support for scientific theories? In *Applications of Inductive Logic*, ed. L. J. Cohen and M. Hesse. Oxford.
Niiniluoto, I. 1978. Discussion: Dretske on laws of nature, *Philosophy of Science*, **45**, 431.
Niiniluoto, I. and Tuomela, R. 1973. *Theoretical Concepts and Hypothetico-Inductive Inference*. Dordrecht.
Pietarinen, J. 1972. *Lawlikeness, Analogy, and Inductive Logic*. Amsterdam.
Quine, W. V. O. 1969. *Ontological Relativity and Other Essays*. London.
Suppes, P. 1966. Concept formation and Bayesian decisions. In *Aspects of Inductive Logic*, ed. J. Hintikka and P. Suppes, p. 41. Amsterdam.

6 Necessities and universals in natural laws

D. H. MELLOR

1 *Prologue.* How do laws of nature differ from cosmic coincidences? This is a question very familiar to philosophers of science, and answers of two sorts still vie for their allegiance. One sort locates the difference in what laws say, the other "in the different roles which they play in our thinking", as Braithwaite's *Scientific Explanation* put it (1953: 295). In Chapter 9 of that book, Braithwaite developed and defended a classic answer of the second sort: the difference, he says there, lies in why we believe laws, not in what they say. In the quarter century since then, other answers of the same sort have been devised: Hesse presents one in this volume. But since then also, answers of the first sort have again come into fashion. The revived fashion has mostly been for reading laws as saying how things must be; but some, more recently, have read them instead as relating not things but properties of things to each other. Hesse notes these fashions and rejects them, to my mind rightly, but she does not elaborate her reasons. It seems to me therefore that I can best complement her article by inspecting these fashions' argumentative cut, to see if they do indeed fit better than her and Braithwaite's Humean gear. Only first I shall build the problem up in my own way, to provide a lay figure to hang the garments on.

2 *The problem.* Certified laws of nature are the primary products of scientific thought and observation. They embody the generalized knowledge which science yields; they supply explanations and predictions of events; and they underlie the design of most modern

Note: This article was written during my tenure of a Radcliffe Fellowship, for which I am indebted to the Radcliffe Trust.

artefacts. To take just three obvious examples: our human life has been much altered in this century by the discovery and applications of laws governing plant genetics, aerodynamics and electromagnetic radiation.

Laws differ widely in their subject matter, importance and complexity. What they have in common is generality. A law says that *all* things or events of some kind have a certain property or are related in a certain way to something else. If the law is statistical, the property is having a chance of having some other property or of being related to something else. It is, for example, a law that all light has the property of going at the same speed in a vacuum; and it is a statistical law that all atoms of the most common isotope of radium have the same chance (fifty–fifty) of turning into something else within their half life of 1622 years.

What needs certifying about a law is its truth. We cannot know that all light goes at the same speed in a vacuum unless it truly does so. Its constant speed will not serve to explain or predict anything if its speed is not in fact constant. And it is unsafe to base the design of artefacts on what is not the case. We know of course that even a certified law may turn out to be false. But without good reason to think it true, we lack good reason to employ it as we do. This is why we do not call something a law unless we think it true, so that a false generalization cannot be a law, although it may be "lawlike": *i.e.* such that it would be a law if only it were true.

Certifying the truth of some laws presents no problem. These are the analytic laws, those whose truth follows from the meanings of the terms they are couched in. There are more reasons than one for laws being analytic. A law may be analytic because it is used to define one of its terms. Newton's laws of motion, for example, may well be analytic because between them they define the Newtonian concepts of force and mass. Or a law, not originally analytic, may become so successful and theoretically important that its terms change their meaning to make it analytic. For this reason it is now arguably analytic that light is electromagnetic radiation, although that could not have been the case when the electromagnetic theory was first conjectured to apply to light. Then, we could easily have envisaged observing light to go faster or slower, for example, than the theory can be shown (by measuring the ratio of electromagnetic to electrostatic units) to require electromagnetic radiation to go. Nowadays we should take such an observation to show some error in the theory rather than question the law that light is electromagnetic radiation.

But even if some laws are analytic, most laws are not, and these are the ones that concern me. It does not follow from the meanings of the terms involved that radium's half life is 1622 years, nor that benzene is as insoluble in water as it is. Nothing semantic prevents a little more benzene sometimes dissolving in water, or some piece of radium having a rather different half life. How then can we certify the truth of what the law says, namely that these things never happen? We cannot see that they never do, if only because at no time can we see that they never will do in the future. We cannot directly perceive the truth of nonanalytic laws. At most, our senses can show us some of a law's past instances, and then only instances of laws about relatively observable properties of things and events. We can observe the speed of this or that ray of light and, perhaps indirectly, the half life of this or that piece of radium; but not all the things and events, past, present and to come, to which the law applies.

The problem then arises why a supposed law should be expected to hold in instances as yet unobserved; in short, Hume's problem of induction. Unlike Popper and his followers, I believe that induction does present a genuine and serious problem, which needs solution and has not yet been solved; although I believe Braithwaite's (1953: Ch. 8) attempted solution is along the right lines. But wherever its solution lies, Hume's problem does not arise only incidentally for laws of nature. On the contrary, it is an inevitable concomitant to their role in supplying predictions. To make a prediction is to anticipate, rightly or wrongly, the result of making an observation; to say or just to expect, for example, that a bomb will explode before we see it do so. Whatever purports, as a law does, to justify such an expectation necessarily arouses Hume's problem. Only a generalization certified by observing all its instances would be free of inductive pretensions, and such a generalization is not much use for predicting things. It might indeed have some use: one might accept it on someone else's authority and use it to predict some instances one had not observed oneself. But real laws are used amongst other things to predict the results of future observations, and these are not yet available to anyone to certify the law with (see Mellor 1979). Real laws therefore undeniably need inductive support.

The other philosophical problem which laws of nature present is the one that concerns us. It is less obvious than the problem of induction, but perhaps more tractable: what exactly do laws say? I have taken them to be generalizations, and there is not much doubt of that. The debatable question is whether laws are more than generalizations, and if so, what more. Now if giving laws one content rather than another made the problem of induction soluble for them, this would be a strong argument for giving them that content. But since I believe no such solution is presently available for any credible content, I must look to other arguments. Hume's problem does, however, provide a reason for preferring weak readings of natural laws. The less a law says, the less there is to be certified in claiming it to be true.

The weakest reading seems to be the obvious one I have already given:

(1)   All *F*s are *G*s,

where *F* and *G* are properties of things or events. They may be relational, comparative or quantitative properties; in statistical laws *G* will be some determinate chance of having another property. (1) is of course a very simple form of law, but it will do; it has all the relevantly problematic features of more complex forms. But before discussing its supposed deficiencies, some preliminary points need to be made clear.

First, as my examples have already illustrated, the 'are' in (1) is to be taken tenselessly. The law applies to all *F* items in the universe, past and future as well as present. The laws of radioactivity do not just give radium's present half life; they say what it always was and always will be. Now some of what we take to be physical constants, such as the half life of radioelements, might indeed turn out to depend on the age of the universe. But then the true laws of radioactivity would say what the dependence was. Those laws would, like all other true laws, apply at all times; the values of our supposed constants at particular epochs being merely special cases of the general laws.

Secondly, I take it that anything in the universe is definitely either *F* or not *F*, either *G* or not *G*. This is not an uncontentious claim. Some have been led to deny it of so-called "vague" properties like being bald, because of its seemingly absurd consequences (for example, that at some point adding just one hair to a bald man's head gets rid of his baldness). Others have been led to deny it of some things and events in the future, either because they want the future to be open, at least in some respects, to being made definite by human decision and consequent human action or because of problems raised by quantum mechanics. They think it cannot now be the case, for example, that I shall definitely either be dead or be alive next year, if it is still open to me and others to settle the matter by what we decide to do between now and then. I think that these are both inadequate grounds for denying that everything is definitely *F* or not *F*, but I shall not argue the point here. (On the first, see Cargile 1969; on the second see Mellor 1981. I also think my being wrong in either case would make little difference to the ensuing discussion, but I shall not argue that either.)

Thirdly, I exclude from the range of *F* and *G* factitious properties such as Goodman's (1965) notorious "grue" (= green if the item is inspected before a specified time, otherwise = blue). I hope and believe criteria can be given to rule out these phoney properties (see for example Hesse 1974: Ch. 3); but in any event all parties agree that they are phoney, and I shall take their exclusion for granted.

I should however emphasize that I do not mean to restrict *F* and *G* to physical, as opposed to psychological or social, properties. Some philosophers (*e.g.* Davidson 1970; McGinn 1978) deny the existence of laws relating nonphysical properties; but largely because they mistake laws to involve necessities of the kind I shall be concerned to dispute and which they correctly perceive to be absent from mental and social generalizations. Anyway the point should be left open here; so if I stick to physical examples, it is only to avoid irrelevant controversy, and not because I think there are no others.

With this preamble, we may now ask what, if anything, is wrong with (1) as a reading of laws of nature. To see what seems to be wrong, we must look at (1)'s consequences in special cases, particularly the case, on which Braithwaite concentrates, where nothing in the world is *F*.

One might imagine that it did not matter what follows from (1) when nothing is *F*, but it does. Let us call a law 'vacuous' in that case. Many important laws are vacuous in this sense. The most famous one is Newton's first law of motion, that bodies acted on by no forces are at rest or move at a constant speed in a straight line. The law is central to Newtonian mechanics, but Newton's own gravitational theory implies its vacuity, since the theory says that all bodies exert gravitational forces on each other. No doubt Newton's laws of motion are peculiar, since as already remarked they may well be analytic. But Newton's first law illustrates a vacuity which is shared by many laws that are in no way analytic. There is in particular a multitude of nonanalytic laws quantifying over determinate values of continuously variable determinables: for example, the laws relating the vapour pressure of substances to their temperature. Each determinate value of these determinables yields another law as a special case, such as the law giving the boiling point of water at atmospheric pressure. Now there are infinitely many different temperatures and pressures, and hence infinitely many of these derived laws, all with mutually incompatible antecedents (nothing can be wholly at two different temperatures or pressures at the same time). Although the temperature and pressure of any given mass of water will vary continuously with time, there are many temperatures which no mass of water ever reaches: temperatures, for example, so high that water would decompose before it reached them. At any rate, so far as these derived laws are concerned, it is entirely accidental whether any water ever is at the temperatures and pressures they apply to. Consequently they must certainly be so construed as to make equal sense whether they happen to be vacuous or not (*cf.* Ayer 1956: 224–5).

In particular, it seems obvious that mere vacuity should not settle the truth of a law regardless of its content. But a lack of Fs makes 'All Fs are As' true for any A, including both $A = G$ and $A = $ not-G. If there never is any water at some temperature T, statements crediting all water at that temperature with any pressure whatever all come out true. That seems absurd; so vacuous laws should be read as saying something other than 'All Fs are Gs'. The question is what.

The obvious answer is that a vacuous law says

(2)   If there were Fs, they would be Gs.

But there are objections to (2). One is that it appears to imply that there are no Fs, whereas laws, even if they happen to be vacuous, certainly do not claim to be. We could in reply say that (2) is not to be read as having this implication; and this stipulation can indeed be given some independent rationale. A case can be made for saying that the implication is not part of what (2) says, but follows rather from applying general rules of discourse: namely, not to mislead, and to be as informative as possible (see Mackie 1973: 75–7). These rules dictate that one should not say (1) when the law is known to be vacuous, since (1) is no more true then than is any other generalization starting 'All *F*s are . . .'. To pick out as a law the generalization which relates *F* especially to *G* in these circumstances, one must have some reason other than its truth. The reason may not be specified, but the fact that there is one is signalled by using (2) instead of (1). Consequently, even if the law says no more than (1), (2) would normally be used when, but only when the law is known to be vacuous. So (2) will indeed signal its user's knowledge of the law's vacuity, even though that is no part of what (2) is being used to say.

This is one of the arguments which can be used to defend Humean accounts of laws as saying no more than (1). It still leaves the problems of saying what reason there is to link *F* and *G* as a law does when there are no *F*s, and why (2) should be the right way to signal this reason. These are among the problems that have exercised Braithwaite and his Humean successors. But since my concern here is with their rivals, I shall concentrate instead on recent attempts to solve the problem of vacuous laws by giving (2) some assertible content over and above (1).

Laws, I have remarked, do not claim to be vacuous, even if they are; and ideally, they should say the same thing whether they are vacuous or not. It will hardly do to make laws say (2) if they are vacuous and (1) if they are not. A law cannot say (1) in both cases, we are supposing; can it say (2)? We have dealt with the obvious objection by removing (2)'s counterfactual implication (that there are no *F*s), which would have made all nonvacuous lawlike generalizations false regardless of their content. What can be said positively in favour of the suggestion?

Consider the universe of non-*F* things or events of which a vacuous law says that if they were *F*s they would be *G*s. It is surely immaterial to this supposed fact about these things or events that there happens to be nothing else which is *F*. So perhaps we should take the *non*vacuous law also to say of every non-*F* thing or event that if it were *F* it would be *G*. But again, the law itself does not assert that these things or events are not *F*. It should say the same of all things or events, whether they are *F* or not. Let us therefore take a law to say of every thing or event *x* that

(3) If *x* were *F* it would be *G*.

(Those who believe in possible as well as actual things and events may take 'x' to range over them too.) I shall take the problem for our non-Humeans to be that of saying what (3) means in this case.

I shall not demand of them a general analysis of so-called 'subjunctive' or 'counterfactual' conditionals like (3). A general analysis would of course have to cover those that we are supposing to give the content of natural laws. But I am not convinced that other uses of these conditionals are homogeneous enough with this one to shed much light on it. In most other uses, for example, (3) might very well imply that x is not F, which we have seen it cannot do here. Or again, to make (3) true of an x, it may often suffice for that x to be F and also G. Lewis's influential analysis, for example, takes this more or less for granted, and the way he reluctantly accommodates possible exceptions (1973: 29) will certainly not cope with natural laws. Yet natural laws must be exceptions: it might be a coincidence that an x is both F and G, and not a matter of natural law at all. So in this case it must take more than that to make (3) true of any x. And as our consideration of vacuous laws has shown, the extra cannot be that all other Fs are Gs too, for there might just as well be no other Fs. So whether the law is vacuous or not, the truth of (1) will not suffice to make (3) true of everything. But what more than (1) can a law say?

3 *Possible worlds.* The traditional non-Humean answer is that natural laws are or express necessities of some kind: what makes (3) true of everything is that Fs not merely are Gs, they have to be; (1) is not merely true, it is necessarily so. Conceptions of law as what Kneale (1949) called 'principles of necessitation' are of course by no means new. The problem with them is to justify the idea of necessity they invoke and to show how it explains the universal truth of (3). Of late years, the development of so-called "possible world semantics" has made that problem look more tractable, and thus encouraged a revival of the idea that natural laws are necessary truths. It has done this by providing a systematic way of saying what makes statements of necessity (and of possibility) true. So in particular we might hope to find in it an acceptable way of saying what makes necessary natural laws true.

The basic concept of this semantics is that of a possible world. A possible world is a way the world might be, or might have been. There are many such ways, and therefore many possible worlds, of which the actual world is just one. Possible worlds are distinguished by what the facts are supposed to be in them: if the supposed facts differ at all, so do the worlds. I might, for instance, die in various ways, and, for each way, at various ages. So there are many possible worlds in which I expire of, say, cirrhosis (or my counterpart in that world does so; see Lewis 1973: 39), and these differ amongst other things according to my or my counterpart's age at the time. In general, a statement which might be true, but fails to specify every detail of the universe, will be true in many possible worlds, differing amongst themselves in the details left unspecified.

Having in some such manner as this grasped the idea of possible worlds, and reified them, one can turn round and give, as the truth conditions of a statement, the set of possible worlds in which it is true. That is how possible world semantics offers to give the meaning of various kinds of modal statements, and in particular of statements of necessity and possibility. How enlightening this conceptual round trip is, from what might be the case, to what is the case in a possible world, and back again, is a very moot point, but one that can be waived while we see how well the concept copes with the supposed necessity of natural laws.

It follows at once from the definition of a possible world that a statement which might be true is one that is true in some possible world. Hence statements which have to be true are those which are true in all possible worlds. In particular, for (1) to be necessarily true is for it to hold in all the worlds there might be or might have been. Is that really what a natural law claims?

Suppose it is: does that solve the problem of vacuous laws and explain (3)'s being true of everything in the actual world? Consider again the case where there are no actual *F*s. The law does not say there are none, and it is tempting to suppose there always might have been. If that were so, then, on this view of laws, (1) would have to be true not only in this world, but also in worlds containing *F*s where its truth would not be the trivial consequence of vacuity it is here. And that would certainly distinguish (1) as a law from other vacuously true generalizations.

But this account depends on the possibility of there being *F*s; and, on this view of laws, there will often be no such possibility. I have cited the example of high temperature instances of the vapour pressure law for water that are vacuous because water decomposes before it reaches those temperatures. Now, that water decomposes below these temperatures is itself a natural law and so, on this view of them, necessary. Consequently these high temperature instances of the vapour pressure law not only are vacuous, they have to be. There could be no water at such temperatures. But that is to say there are no possible worlds in which these instances are not vacuous; and therefore none in which the truth of this instance of (1) is other than a trivial consequence of vacuity.

So the idea of laws being true in all possible worlds does not solve the problem of vacuous laws. Nor, for much the same reason, does it explain why (3) is true of everything in this world. Again, it would if anything, a, in this world might have been F even if it isn't. Then there would be possible worlds in which a (or some counterpart of a) is in fact F; and in all these worlds it, like every other F, is G. Where that is so, it seems to me undeniable that (3) is true of a. However, for any F there will be many as of which it is quite incredible that they might have been F. Take the law that in a vacuum all light goes at a constant speed – which is to say that all photons do. It is true then, of anything at all, that it would go at that speed in a vacuum if it were a photon. But this is not to say of everything that it might have been a photon. There is no possible world in which I am (or any counterpart of me is) a photon; and a fortiori none in which, as a photon, I (or any counterparts of me) travel at the speed of light. That is not, I believe, what makes this instance of (3) true of me. Yet I believe it is true of me, since I believe the law; and there is surely no inconsistency in my combining these beliefs.

For subjunctive conditionals like (3) to be true, their antecedents do not have to be possible. This is blatantly obvious in *reductio ad absurdum* proofs, where the truth of a subjunctive conditional is actually used to prove that its antecedent is *not* possible. One and one cannot make three precisely because, if they were to, something impossible would be the case. It should be almost as obvious that conditionals which give the content of natural laws likewise do not imply the possibility of their antecedents being true. The vapour pressure example shows at least that they cannot both do this and themselves be necessary truths. And I have given elsewhere (1974: 173) the example of safety precautions at a nuclear power station, which are supposed to make impossible the conditions under which, as a matter of natural law, the fuel would explode. It is ridiculous to maintain that the success of these precautions would disprove the very law that makes them necessary.

I am not sure why (3) should be so often thought to imply that *x* might be *F*. The reason may well be the same for taking (3) to imply that *x* is not *F*: namely, that it is customary to reserve subjunctive conditionals for use when their antecedents are believed to be false but possible. We see, however, that this custom is not invariable, and have in any case seen reason (see p. 110) not to make such a custom part of a conditional's meaning. So however natural the thought may be, it is mistaken, at least of the conditionals implied by natural laws. But the mistake is very widespread and of long standing, and it has had serious consequences. It has bedevilled the analysis of disposition statements, as I argued in §9 of my (1974). It has likewise afflicted discussions of free will, in which 'I could have done *X*' is frequently equated with 'I would (or could) have done *X* had I chosen to'. But it obviously does not follow from the latter that I could have done *X*, since it obviously does not follow that I could have chosen to.

The common confounding of conditional statements with statements of possibility has thus had ill effects in more than one area of philosophy. The ill effect here has been that possible world semantics have been mistakenly thought to give sense to the idea that natural laws are, or assert, some kind of necessity.

4 *Natural necessity*. Laws might however still be necessary even if possible world semantics fails to say what makes them so. What makes (3) true of everything might still be that nothing could be both *F* and not *G*, whether or not it could be *F*. But it is not at all obvious that this is so. Subjunctive conditionals are not in general made true by necessities. Suppose that if I were to go to London I would go by train. This does not mean that I could not go any other way, merely that I would not. Lewis's (1973) treatment of subjunctive conditionals recognizes this fact about them: the consequent does not have to be true in all the possible worlds the antecedent is true in, only in those most like the real world.

Still, I have insisted that conditionals like (3) which follow from laws are a special case. In particular, it does not suffice for their truth that their antecedents and consequents are true; whereas my going to London by train may well make it true that, were I to go, I would go that way. So perhaps (3) does need some necessity to make it true of everything, even if conditionals in general do not.

But most natural laws seem to be contingent. Apart from those that are definitions, and those whose success has made them analytic, any law might have been false. We could have come across a counter-example to it; and we still could, even if we never will or would. That seems at any rate to be why we need to test our supposed laws by observation: things could be other than the law says, so we need to look and see whether or not they are. I believe, for example, that light could have gone in a vacuum at other than its constant speed, even if no photon ever does and even if nothing, were it a photon, ever would. So on the face of it, conditionals like (3) no more exhibit necessity than does the conditional about my going to London by train.

Attempts have been made to explain away the apparent contingency of natural laws. One attempt, which need not detain us long, distinguishes logical necessity and possibility from their natural or physical counterparts. It is logically possible for Fs not to be G, but not naturally or physically possible. But all 'physically possible' means is 'consistent with natural law'. So to say that something is physically necessary is merely to say that some law entails it. Whether the law says it has to happen, and whether the law itself has to be true, remain entirely open questions.

A more serious attempt distinguishes between metaphysical and epistemic necessities (Kripke 1971: 150–1; Dummett 1973: 121); that is, between being necessary and being knowable *a priori*. Laws appear to be contingent because they cannot be known *a priori*. They cannot be proved in the way the truths of logic and mathematics can. We need to look and see what the world's laws are, and it may always turn out that what was thought to be a law really is not one. The Fs we have seen to be G may mislead us into believing they all are, even though some future ones are not. It is consistent with all we have seen that there should be Fs which are not G. That is the epistemic possibility of a supposed law being false; and something like it exists in mathematics. There too, special cases may mislead us into believing a mathematical generalization to which there are in fact counterexamples. Now, recognizing this possibility in mathematics does not diminish our belief in the necessity of mathematical truths: if the generalization *is* true, it could not have been otherwise. It is likewise conceivable that natural laws, if true, are necessarily so, even though we may be mistaken in what we suppose the true laws to be.

The apparent contingency of natural laws could undoubtedly be explained away like this if there were good reason to think them necessary: but is there? The analogy with mathematics certainly does not give one. If Goldbach's conjecture proves true, any attempt to suppose it false will eventually lead to contradiction (that of course being one way of proving it). In that case no consistent description could be given of a world in which the conjecture was true. That is, there is no such possible world. We might therefore explain the conjecture's necessity, if true, as truth in all possible, *i.e.* coherently conceivable worlds; since conceivability is a notion arguably more basic than necessity and intelligible independently of it. But no such case can be made for the corresponding conception of natural necessity. As Hume insisted, there is no difficulty in conceiving a natural law to be false: since it is not analytic, no contradiction ensues. A perfectly coherent description can be given of a world containing *F*s that are not *G*. The only ground for thinking such a world impossible would be that the law which would be false in it is not only true but necessary; and this is the very fact that needs to be established and explained.

5   *Essences*. Arguments have recently appeared for the metaphysical necessity of some laws, namely those specifying essential properties of natural kinds. An essential property of a kind is one which nothing of that kind can lack. So if being *G* is of the essence of a kind *F*, the law that all *F*s are *G*s will be a necessary truth. The exemplars most widely touted by advocates of essences concern the microstructure of kinds: the atomic number of gold, the molecular constitution of water, the genetic makeup of plant and animal species, and the mean kinetic energy of gas particles at a given temperature. The question is: why suppose that these, or any other, properties of kinds are essential?

Two sorts of arguments have lately been adduced for essences, and hence for the necessity of the corresponding laws. Both employ possible world semantics; neither therefore proves more than that some generalizations hold in all possible worlds, and we have seen in **3** that this is not enough to serve our turn. But the arguments repay scrutiny nonetheless, since there is more to them than the possible world jargon they are couched in.

One argument, due to Putnam (1975), infers essences from a mechanism for fixing what things or events a kind predicate ('*F*') applies to, *i.e.* its extension. This mechanism fixes what things are,

or might be, Fs in two stages. First, there are archetypal actual Fs
(*e.g.* paradigm specimens of gold or water): things that have to be F
if anything is. Second, to be F, anything else has to have a suitable
'same-kind' relation to these archetypes. What this means is that it
has to share some property with them – apart of course from the
property F. What the same-kind relation is, for any category of
kinds, it is for empirically testable scientific theories to say. The
relations are not discoverable *a priori*, and in particular they do not
follow from the meanings of the predicates involved: the laws
giving the essences of kinds are not supposed to be analytic. But any
shared property G which a same-kind relation picks out will be an
essential property of the kind since, Putnam assumes (1975: 232),
the relation is an equivalence relation holding across all possible
worlds. Thus not only are actual Fs all Gs, all possible Fs are: so (1)
in this instance is true in all possible worlds.

I have elaborated elsewhere (1977) my reasons for rejecting this
argument. Briefly, the extensions of real natural kinds do not in the
first place depend on archetypes in the way Putnam's mechanism
requires. And, in the second place, even if they did, his mechanism
would still not produce essences. To produce an essence, the same-
kind relation must be transitive, in order to ensure that all possible
Fs share the *same* property G with each other. But the mechanism
does not need a transitive relation, since what makes things Fs in
other possible worlds is their sharing some property other than F
with the archetypal Fs in this one, and there is nothing to say this
shared property must be the same in every possible world. For
Putnam to claim the same-kind relation to be transitive, which he
does in taking it to be an equivalence relation, is for him gratuit-
ously to assume the essentialist conclusion he is out to prove. His
mechanism in fact gives us no reason to think any instance of (1)
true in all possible worlds. And since in any case only those giving
essences are in question, Putnam's theory, even if it worked, would
not solve the general problem of distinguishing laws from universal
coincidences.

The same of course is true of Kripke's (1971, 1972) argument for
essences; but that too we must look at, since solving our problem
for some laws would at least be better than solving it for none.
Kripke's argument is quite different from Putnam's. Kripke takes
laws giving essences to be identity statements: the law 'Water is
H_2O' he takes to say not merely that anything, were it water, would

be H_2O, but that being water and being H_2O are one and the same property. But identity is a necessary relation, in the sense that nothing could fail to be identical to itself. So being water and being H_2O are the same property in all possible worlds, not only in this one. Nothing that could be water could fail, were it water, to be H_2O.

This of course is the merest sketch of Kripke's argument. He has, for example, also to show that 'water' and 'H_2O' are what he calls 'rigid designators', *i.e.* that they refer to the same stuff in any possible world it exists in. Otherwise the identity statement, since it is not analytic, might be true without being necessary, as for example 'Water is the most powerful solvent' is. ('The most powerful solvent' is not a rigid designator: it refers to whatever the most powerful solvent is, which in a world restricted largely to oil products would not be water.) As I have abbreviated Kripke's argument, so I shall abbreviate the objections raised in my (1977). The chief objection is that the argument, like Putnam's, blatantly begs the question: for being water and being H_2O to be the same property at all, never mind necessarily, the predicates '. . . is water' and '. . . is H_2O' must already be coextensive in all possible worlds. This is not a conclusion to be derived from the necessity of the identity: it is built into the identity as a premiss. Granted, 'Water is H_2O' states a true law, and it has the form of an identity statement. But it is clearly only a variant of 'All water is H_2O', which does not have that form. At any rate, the identity of these properties only follows if 'water' and 'H_2O' are rigid designators, *i.e.* could not refer to different properties. But to believe this, one needs already to believe what the argument from this premiss is supposed to show: namely, that there could not have been some samples of water of a different molecular constitution.

Kripke, like Putnam, fails to establish the existence of essences. The microstructural exemplars which give their doctrine its spurious appeal indeed have a special status in science, but not the status of essences. They are special because they are central to our current scientific theories; but that, I have argued elsewhere (1977: §7), is quite a different matter from being necessary features of the world.

6 *Universals.* The properties of being water and being H_2O do not stand in the necessary relation of identity. Perhaps, however, as Armstrong (1978a: Ch. 24), Dretske (1977) and Tooley (1977) have suggested, these universals stand in some contingent relation which makes it a law that all water is H_2O. This relation, that is to hold between the properties F and G whenever 'All F are G' is a law, Armstrong and Tooley call 'nomic necessitation'; I shall call it 'N'. F and G have to be differently related if the law is that no F are G or that all F have a chance p of being G. But N will do for now: if it works, the other relations will; if not, nor will they.

This suggestion requires a realist view of at least those universals which are related by natural laws: for N to relate F and G, these properties must exist. That of course is debatable, but suppose for the moment it is true. Then FNG is by definition the fact that makes 'All Fs are Gs' a law. This is a contingent fact, and not only because F and G might not exist. F and G could quite well exist without 'All Fs are Gs' being a law: laws do not relate every property to every other property. Being water and being at 100°C, for example, are properties that enter into laws, yet no law relates them to each other. But though it is not, it might have been a law that all water is at 100°C; just as it might not have been a law that all water boils at that temperature at atmospheric pressure. Apart from analytic laws, therefore, it is quite contingent that N relates any particular F and G.

To do its job, N has not only to make Gs out of actual Fs, it has to make (3) true of everything, *i.e.* to be such that anything, were it F, would be G. Since this is all it has to do and be, one might think that postulating N is more a relabelling of the problem than a solution to it. But that would not be a fair response. There is a dearth of candidates for making (3) universally true, as we take it to be. If F, G and N would between them make it true, that may well, as Tooley (1977: 262) urges, be reason enough to believe in them.

After all, we already invoke properties to make conditionals true. The inertial mass, m kilogrammes, of a thing a at time t makes true all the conditionals of the form

(4)   If a were subjected at t to a force of f newtons, it would then
        accelerate in the direction of the force at f/m metres/second².

Any of these conditionals is in fact a generalization about events, namely that, if they were subjectings of a to a (specific) force f at t, they would be (or be shortly followed by) accelerations of a of magnitude f/m. These generalizations are just like the conditionals (3) entailed by laws, except that they are restricted to the individual a. We think them true, and a fact is needed to make them so: and the requisite fact is that *a* has mass *m* at time *t*. This is all the property of inertial mass amounts to: a truth-maker, as Tooley puts it, for conditionals like (4); and I have argued elsewhere (1974) that all properties of things are just truth-makers for such conditionals. But if we believe in properties *F* and *G* because they are needed (and suffice) to make conditionals like (4) true, why jib at accepting N when it is likewise needed and (with *F* and *G*) suffices to make conditionals like (3) true?

Here, however, the crucial difference between (3) and (4) emerges: (4) entails that *a* exists. Without *a*, '*a*' would have no reference, and I do not see what (4) could then mean, nor how in particular it could be true. So if (4) is true, *a* exists; so the fact that *a* has mass *m* is always available to make (4) true. But it is by no means so clear that *F* and *G* exist whenever (3) is true of everything. The law that all *F* are *G*, it is agreed, may be vacuous: and if it is, there are no *F*s. Now if, as many (including Armstrong) suppose, properties and other universals need instances, then without *F*s there will be no *F*. But without *F* there will be no fact *FNG* to make (3) true of everything; and the problem of accounting for vacuous laws will remain unsolved.

Perhaps then universals need no instances. Concepts certainly do not: we can have the concept of a unicorn without there being unicorns. But universals are not concepts: concepts, if anything, are parts of our thought or our language; whereas universals, if anything, are parts of the world whether or not it contains any thought or language or concepts. No doubt concepts are closely related to universals, but it is not safe to assume that universals can dispense with instances just because concepts can. That remains an open question.

Tooley takes it for granted that universals need no instances, since he uses a particular example of a vacuous law to argue by elimination "that it must be facts about universals that serve as the truth-makers for basic laws without positive instances" (1977: 672), going on to ask rhetorically: 'if facts about universals constitute the truth-makers for some laws, why shouldn't they constitute the truth-makers for all laws?' Armstrong, by contrast, holds a "Principle of Instantiation: For each N-adic universal, U, there exist at least N particulars such that they U" (1978: 113); but he offers nonetheless to cope with Tooley's example of a vacuous law. Now if Armstrong really can supply enough universals to make vacuous laws true, without violating his principle of instantiation, we may not have to decide whether universals do in fact need instances; but can he?

In Tooley's example, as Armstrong puts it, just two out of several types of particle happen never to meet; so the law governing their interaction is vacuous. Nevertheless particles of other types meet, so that the universal, meeting (M), exists; as do these two mutually evasive particle types (A and J). Armstrong can claim therefore that the law, despite being vacuous, 'holds in virtue of the universals [A, J, M] being what they are' (1978a: 157). This solution, however, is only available for special cases of vacuous laws. For a start, it only works here because particles of other types do meet, thereby ensuring the existence of the universal M. Now if the law governing A and J particle interactions does not ensure their meeting, it can hardly ensure the meeting of other types of particles; and if A and J particles can fail to meet, so can others. If no particles ever met, the laws of all their interactions would still be true, but the universal M would not, for Armstrong, exist to make them so.

More seriously, not only might there be no meetings, there might be no A or no J type particles. Yet the law could still be true and important, even if there were nothing it applied to. We have seen that to be the case with Newton's first law of motion, and the vacuous instances of vapour pressure laws. For these, and indeed for the bulk of vacuous laws, Armstrong's principle of instantiation does deprive him of the universals he needs as truth-makers. The Tooley case he discusses happens to be of the only sort he can cope with, and it is worth drawing out what makes cases of this sort amenable to Armstrong's treatment: namely, that in them there exist things with properties which make certain generalizations true. These are in fact generalizations of conditionals like (4) above. Consider that for many determinate values of the determinable force f, (4) is vacuous: a can only be subjected to one (net) force at any one time, and there will be many forces a never experiences. Yet a's always having mass m suffices to make all these vacuous generalizations true. And so it is with Armstrong's A and J particles. While they exist, they are disposed to interact as the law says, whether they ever actually meet or not. Armstrong's universals A and J are just conjunctions of such dispositions (see Mellor 1974), and can thus be truth-makers for those laws whose vacuity results merely from the dispositions of actual things failing to display themselves.

But not for the more important cases in which laws are. made vacuous by the non-existence of the things themselves; and hereafter I will reserve the term 'vacuous' for such cases.

Since vacuous laws will in fact defeat the Armstrong–Dretske–Tooley account if universals need instances, we have after all to consider whether they do. I follow Ramsey in taking particulars and universals to be simply parts of facts picked out in order to generalize. For example, "It is not '*aRb*' but '$(x).xRb$' which makes *Rb* prominent. In writing '$(x).xRb$' we use the expression '*Rb*' to collect together the set of propositions *xRb* which we want to assert to be true" (Ramsey 1925: 28–9). To recover a proposition from this set, we need to know what an instance of *xRb* is, *i.e.* we need criteria for identifying the items, such as *a*, which have been quantified over. But we do not need separate criteria to identify *Rb*. Given *a*, *Rb* is just the remainder of the fact *aRb*. If it were an independently identifiable constituent, then, as Ramsey says (1925: 23–4), *a*(*Rb*) would differ from (*aR*)*b* and *a*(*R*)*b*, because these facts would have different constituents; and this is absurd.

Similarly, if we form the doubly general '$(x)(y).xRy$', we must regard the universal *R* as just the common part of all the facts thus collected: the fact *aRb* minus *a* and *b*. And since in forming the law that all *F*s are *G*s we at least collect whatever facts such as *Fa* there may be, the universal *F* must likewise just be the common part of all such facts. At least in laws, therefore, a universal must be regarded as derived from the particulars which are its instances and the facts that they are so. To regard them, as extreme realists do, as a primitive kind of entity, distinct from particulars but able to combine with them to yield facts, is to put the universal cart before the factual horse. It does nothing but pose such ancient but manifestly dotty conundrums as: why there are these two different kinds of entity, particulars and universals, and what the difference between them is; why two entities of the same kind (two particulars or two universals) cannot combine to form a fact; what the relation is between a particular and a universal that are so combined. The last question on its own is fatal to this view, since any answer to it immediately generates Bradley's (1897: Ch. 3) notoriously vicious regress; a regress not avoided just by Armstrong's ingenuous device of calling the relation in question a "union . . . closer than relation" (1978a: 3).

The fact is, as Ramsey showed, that we have no *a priori* reason to suppose that universals are fundamentally different in kind from particulars. What we think of as particulars are merely the kinds of entity we can most readily individuate, typically by appeal to their spatio-temporal location (*cf.* Braithwaite 1926); and a universal is just the common residue of a set of facts about such individuals. So there is really no mystery about what relates particular to universal in a fact, nor about why a fact has to contain at least one of each. Nor is there any general reason why residual universals cannot them-selves be individuated and so admitted in their own right as entities to be quantified over – thus, for example, leaving the particular *a* as the common residue of a set of facts about *a*'s properties. Nominal-ism therefore is not the only, nor the most sensible, alternative to an extreme realism about universals.

From this Ramseyan account of universals it does however fol-low that they need instances: Armstrong's principle of instantiation is quite right. In the law that all *F*s are *G*s, the property *F* is just the residue of such facts as *Fa*. If the law is vacuous, there are no such facts; and no facts leave no residue. If there are no *F*s, there is no *F*. So there will be no fact *FNG* to make such a vacuous law true, and the Armstrong–Dretske–Tooley theory fails. Whether there are "real connections of universals", as Ramsey put it, I do not know: like him, "I cannot deny it; for I can understand nothing by such a phrase; what we call causal laws I find to be nothing of the sort" (Ramsey 1929: 148).

7 *Epilogue.* I have considered two attempts, seriously undertaken of late years, to make natural laws say more than generalizations; both fail. The law that all *F*s are *G*s is given the force it needs neither by taking it to say that 'All *F*s are *G*s' is true in all possible worlds or is in some other sense necessary, nor by taking it to assert a contin-gent relation between *F* and *G*. Neither construal can cover the crucial case of vacuous laws which Braithwaite rightly stresses. There are no doubt likewise aspects of the problems of laws which solutions of Braithwaite's Humean cut also have difficulty cover-ing; only they, to my mind, are more readily patched up. Those patches, however, must be woven elsewhere.

University of Cambridge

REFERENCES

Armstrong, D. M. 1978. *Nominalism and Realism*. Cambridge.

Armstrong, D. M. 1978a. *A Theory of Universals*. Cambridge.

Ayer, A. J. 1956. What is a law of nature? In his 1963 *The Concept of a Person and other essays*, pp. 209–34. London.

Bradley, F. H. 1897. *Appearance and Reality*, 2nd edn. Oxford.

Braithwaite, R. B. 1926. Universals and the 'method of analysis'. *Aristotelian Society Supplementary Volume* 6, 27–38.

Braithwaite, R. B. 1953. *Scientific Explanation*. Cambridge.

Cargile, James. 1969. The Sorites paradox. *The British Journal for the Philosophy of Science* 19, 193–202.

Davidson, Donald. 1970. Mental events. In *Experience and Theory*, ed. L. Foster and J. W. Swanson, pp. 79–101. London.

Dretske, Fred I. 1977. Laws of nature. *Philosophy of Science* 44, 248–68.

Dummett, Michael. 1973. *Frege: Philosophy of Language*. London.

Goodman, Nelson. 1965. *Fact, Fiction and Forecast*, 2nd edn. New York.

Hesse, Mary. 1974. *The Structure of Scientific Inference*. London.

Kneale, William. 1949. *Probability and Induction*. Oxford.

Kripke, Saul A. 1971. Identity and necessity. In *Identity and Individuation*, ed. M. K. Munitz, pp. 135–64. New York.

Kripke, Saul A. 1972. Naming and necessity. In *Semantics of Natural Language*, ed. Donald Davidson and Gilbert Harman, pp. 253–355; 763–9. Dordrecht.

Lewis, David. 1973. *Counterfactuals*. Oxford.

McGinn, Colin. 1978. Mental states, natural kinds and psychophysical laws I. *Aristotelian Society Supplementary Volume* 52, 195–220.

Mackie. J. L. 1973. *Truth, Probability and Paradox*. Oxford.

Mellor, D. H. 1974. In defence of dispositions. *Philosophical Review* 83, 157–81.

Mellor, D. H. 1977. Natural kinds. *The British Journal for the Philosophy of Science* 28, 299–312.

Mellor, D. H. 1979. The possibility of prediction. *Proceedings of the British Academy* 65.

Mellor, D. H. 1981. McTaggart, fixity and coming true. *Reduction, Time and Reality*, ed. R. Healey. Cambridge.

Putnam, Hilary. 1975. The meaning of 'meaning'. In his *Mind, Language and Reality*, pp. 215–71. Cambridge.

Ramsey, F. P. 1925. Universals. In his 1978 *Foundations: Essays in Philosophy, Logic, Mathematics, and Economics*, ed. D. H. Mellor, pp. 17–39. London.

Ramsey, F. P. 1929. General propositions and causality. In his 1978 *Foundations: Essays in Philosophy, Logic, Mathematics and Economics*, ed. D. H. Mellor, pp. 133–51. London.

Tooley, Michael. 1977. The nature of laws. *Canadian Journal of Philosophy* 7, 667–98.

7   *Induction as self correcting according to Peirce*

ISAAC LEVI

Richard Braithwaite helped many of us to avoid confusing statistical hypotheses with frequency models for them and taught us thereby to stop exaggerating the importance of so-called "frequency interpretations" of probability beyond their sometimes heuristically useful roles as models.

When Braithwaite wrote *Scientific Explanation*, it was a commonplace that data concerning finite relative frequencies could not be used to establish or refute statistical hypotheses deductively. But the obsession with interpretations of statistical probability in terms of limits of relative frequency which gripped many philosophers in the 1930s, 1940s and 1950s led to neglect of the problem of how to use information concerning finite relative frequencies to render verdicts concerning statistical hypotheses. The question was, for the most part, regarded as a merely "methodological" one best left to statisticians to consider.

Braithwaite was not only free of the obsession but had a correct appreciation of the fundamental significance of the "methodological" problem for the understanding of statistical probability. And he consulted the statistical literature (in particular the ideas of Neyman and Pearson) to devise his own approach to this question.

Charles Peirce insisted on the self corrective character of inductive reasoning. So did Hans Reichenbach. By a doubtful induction, it has been suggested that Peirce is a forerunner of Reichenbach. The inference is illicit and the suggestion false. There is both equivocation and revision of outlook in Peirce's writings; but when Peirce's conception of the self correcting character of inductive reasoning is separated from similar sounding characterizations of scientific method, abduction and deduction, his view is tolerably

127

clear. Peirce's conception anticipates the approach to confidence interval estimation and hypothesis testing later developed by Neyman and Pearson. Like Braithwaite, he thought the "methodological" problem of using data about relative frequencies to reach conclusions about statistical hypotheses to be fundamental to understanding statistical probability and his proposals for dealing with the problem resemble Braithwaite's. The relation between Peirce and Reichenbach is superficial; between Peirce and Braithwaite it is profound.

Two preliminary issues need to be considered prior to explaining what Peirce meant by induction correcting itself. One concerns the relation between induction and abduction; the other concerns the relation between probabilistic or statistical deduction and "necessary" deduction.

In the 'Minute logic' of 1902, Peirce expressed an important reservation about his earlier 'A theory of probable inference' of 1883. The latter paper together with 'The probability of induction' of 1878 will be the main source interpretation of the self correcting character of induction in Peirce's system. It is important, therefore, to understand that Peirce's reservations about that paper in his remarks in 1902 do not concern at all his account of induction in the earlier essay.

Peirce's reservations focused on the distinction between what he called 'Hypothetic Inference' in 1883, 'Abduction' in 1902 and 'Retroduction' or 'Presumption' on other occasions. In 1878 and in 1883, these modes of reasoning are regarded as species of "ampliative" or "synthetic" reasoning in contrast to "explicative", "analytic" or "deductive" reasoning. These subdivisions were based primarily on considerations of form and meaning. An example of hypothetical reasoning is appealing to premisses that all Ms have the characters $P_1, P_2, \ldots, P_n$ and that individual S has the same characters to the conclusion that S is an M. More generally, one might assume that S has r of these n properties and conclude that S has an "r-likeness" to the Ms. In contrast, one argues inductively from the circumstances that n objects have been selected from the Ms and have been found to contain r Ps to the conclusion that the proportion of Ps among the Ms in general is approximately r/n. In the one case, an extrapolation is made from similarities in some respects to similarities in all (relevant) respects. In the other, an extrapolation is made from an observed relative frequency in a sample to a conclusion about relative frequencies in the population.

Even in his earlier work, Peirce acknowledged some difficulty in denying hypothetical inferences the status of inductions. His inclination to do so derived from "the impossibility of simply counting qualities as individual things are counted. Characters have to be weighted rather than counted" (CP.2.707).[1]

In his later work, Peirce drew the distinction between abduction and induction in terms of differences in the tasks they perform in inquiry rather than in terms of formal considerations. Abduction involves the evaluation of hypotheses as worthy of being entertained as potential answers to the question under investigation. Inductive reasoning seeks to render a verdict as to which of the potential answers identified as worthy of consideration at the abductive stage of inquiry ought to be adopted on the basis of the data of experimentation.

According to this classification, neither of the modes of ampliative reasoning recognized in the earlier papers qualify as abductive reasoning. Peirce explicitly acknowledged this in 1902:

By my capital error was a negative one, in not perceiving that, according to my own principles, the reasoning with which I was there dealing could not be the reasoning by which we are led to adopt a hypothesis, although I all but stated as much. But I was too taken up with considering syllogistic forms and the doctrine of logical extension and comprehension, both of which I made more fundamental than they really are. As long as I held that opinion, my conceptions of Abduction necessarily confused two different kinds of reasoning (CP.2.102).

In the 1883 paper on which Peirce was commenting, he asserted that the conclusions reached *via* hypothetic reasoning are usually inaccessible *via* induction and conversely. According to his later view, however, a hypothesis *h* is a potential answer to a given question. At a later stage of inquiry, it may become a settled part of scientific knowledge. Hence, the same proposition may be a hypothesis at one time and knowledge at another. In this sense, the proposition *h* becomes accessible both to hypothetic reasoning and inductive reasoning albeit at different stages of inquiry.

In Peirce's later writings, he distinguished qualitative from quantitative induction and recognized both sorts of reasoning as relevant

[1] References to Peirce are to the *Collected Papers of Charles Sanders Peirce*, 8 vols., Belknap Press. They take the form '(CP.*n.m*)', where *n* is the volume, *m* the paragraph number.

to the inductive task. A comparison of what he says about qualitative induction in his later writing with what he says about hypothetic reasoning in his earlier work suggests that qualitative induction is the reincarnation in his later views of the earlier construal of hypothesis. Quantitative induction, however, corresponds to what he had earlier called induction without any qualification.[2] If this fragment of a translation manual is adopted, it is safe to assume that what Peirce said about induction in 1878 and 1883 continued to be his view in the twentieth century.

Deduction, the third member of Peirce's Holy Trinity, has thus far been ignored. If anything its relation to the other members of the Peircean partition is more difficult to handle than the relations of the other two members to each other. Space limitations excuse me from trying.

But attention must be paid to one type of reasoning which Peirce included under the rubric of deduction. I allude to probabilistic and statistical deduction. These inferences are nowadays called 'statistical syllogisms'. The major premiss asserts that the proportion of Ms that are Ps is p. The minor premiss or "case" asserts that n individuals are selected at random and with replacement from the Ms. The conclusion or "result" is that, with probability k, approximately np of the n individuals sampled are Ps.

In at least one place (footnote 1 to CP.2.719), Peirce explicitly acknowledged an ambiguity in the interpretation of 'with probability k' in the statistical deduction. In one sense, 'with probability k' is part of the conclusion of the statistical deduction. In the other, it indicates "the modality with which this conclusion is drawn and held for true". In that footnote and elsewhere there is considerable evidence that Peirce appreciated the import of the ambiguity and was prepared to exploit two types of statistical deductions.

Probability, according to Peirce, is a property of rules of inference. The probability associated with a rule licensing inference from premises of type A to conclusions of type B is the limit of relative frequency of true conclusions of type B being drawn from true premises of type A in a hypothetically infinite sequence of such inferences starting from true premises. Peirce's formulation of this view was less precise and explicit in his earlier writings than

[2] Peirce says as much in the letter to P. Carus *circa* 1910 (CP.8.233). However, in 8.234, he qualifies this remark. I suspect that the qualification derives from confusion over the import of quantitative determination.

in the first decade of the twentieth century; but there do not appear to have been any philosophically significant changes in his "materialistic" view of probability. In his later writings, Peirce saw fit to emphasize the hypothetical character of the infinite sequence; but there is no evidence to support the view that in his earlier work he had required the existence of an infinite or extremely large but finite sequence for the definability of probability. In this respect, his view is closer to Von Mises' outlook than to Reichenbach's.

Returning to statistical deduction, Peirce sometimes construed the conclusion of a statistical deduction as asserting a claim of statistical probability – *i.e.* of probability defined in terms of limits of relative frequencies in hypothetically infinite series.

Thus, the conclusion asserts that the probability of approximately np of the Ms sampled being Ps on an n-fold random sample with replacement is equal to k. If this kind of trial were to be repeated *ad infinitum* then, in the limit, the relative frequency with which approximately np Ms sampled are Ps would be k.

Peirce contended that the statistical deduction so construed is a deduction or necessary inference in the narrow sense. What he meant is this. To claim that an object is selected at random from the Ms is to assert that it is selected by a method according to which each M has an equal probability of being picked. Hence, if the set of Ms is finite (and Peirce's approach works best when M is finite although he claimed to have extended it to cases where M is infinite),[3] each object in the set has a probability equal to $1/m$ of being selected where m is the number of Ms. If the proportion of Ps among the Ms is p, it follows that the probability (hypothetical limit of relative frequency) of obtaining a P on a random selection of an M is equal to p. Sampling with replacement guarantees independence. Hence, random sampling n times with replacement yields a binomial process and the probability that the relative frequency of Ps among the Ms sampled falls in some interval around p can be calculated and, for a suitable interval, can be shown to be equal to some high value k provided n is sufficiently large. Thus, the conclusion is deduced from the definitions of probability and random selection with the aid of some mathematics.

This reading predominates in CP.2.785. But Peirce's usual practice is to regard the conclusion of the statistical deduction to be the

---

[3] Peirce discussed the question of the cardinality of M in 1883 (CP.2.731) and more interestingly in 1901 (CP.7.209–14).

assertion that the proportion of *P*s among the *M*s sampled is approximately ρ and to regard the assignment of probability *k* as indicating "the modality with which this conclusion is drawn and held for true".

Thus, the conclusion is not itself a statement of statistical probability but is obtained from the premises of the statistical deduction by an inference rule whose associated long run limit of relative frequency of success is deducible from these very same premises.

Whether Peirce's use of 'deduction' is appropriate for this kind of reasoning is a very minor matter. Of greater interest is how Peirce thought the probability *k* indicated the modality in question. Peirce explicitly worried about this issue in 'The doctrine of chances' of 1878:

> According to what has been said, the idea of probability essentially belongs to a kind of inference which is repeated indefinitely. An individual inference must be either true or false, and can show no effect of probability; and, therefore, in reference to a single case considered by itself, probability can have no meaning. Yet if a man had to choose between drawing a card from a pack containing twenty-five red cards and a black one or from a pack containing twenty-five black cards and a red one, and if the drawing of a red card were destined to transport him to eternal felicity and that of a black one to consign him to everlasting woe, it would be folly to deny that he ought to prefer the pack containing the larger proportion of red cards, although, from the nature of the risk, it could not be repeated (CP.2.652).

Peirce then points out that the problem is just as poignant for actuaries, professional gamblers or anyone who takes a large number of risks. In an infinite sequence of gambles where the limit of relative frequency of success is known, one can compute the long run average gain. But no matter how large a finite subsequence might be, there is no guarantee that the success rate approximates the probability on a single trial:

> We are thus landed in the same difficulty as before, and I can see but one solution to it. It seems to me that we are driven to this, that logicality inexorably requires that our interests shall not be limited. They must not stop at our own fate, but must embrace the whole community. This community, again, must not be limited, but must extend to all races of beings with whom we can come into immediate or mediate intellectual relation. It must reach, however vaguely, beyond this geological epoch, beyond all bounds. He who would not sacrifice his own soul to save the whole world is, as it seems to me, illogical in all his inferences collectively. Logic is rooted in the social principle (CP.2.654).

After pointing out that men are not as selfish as advertised, Peirce removes all the teeth from the demand for altruism:

Now it is not necessary for logicality that a man should himself be capable of the heroism of self sacrifice. It is sufficient that he should recognize the possibility of it, should perceive that only that man's inferences who has it are really logical, and should consequently regard his own as being only so far valid as they should be accepted by the hero. So far as he thus refers his inferences to that standard, he becomes identified with such a mind (CP.2.654).

Thus, the person confronting the grisly gamble with the cards need not literally embrace the aims of the altruistic hero. And, as Peirce explicitly states, he need not assume either the existence of the hero or the availability of the opportunity to promote infinitely long run community interests.

On the other hand, the ordinary mortal assessing the particular risky situation confronting him should, if his reasoning is to be logical and valid, simulate the reasoning which an altruistic hero would carry through in seeking to promote community interests. For Peirce, validity is assessed by appeal to an ideal observer who is the altruistic hero of the passages just cited.

Peirce wished to validate what we might call credal, personal or subjective probability judgements under certain special circumstances; but he sought to do so without granting legitimacy to the "conceptual" view of probability he attributed to the Laplacians against whom he argues so vigorously. He tried to achieve his goal by restricting the assignment of numerical expectation-determining probabilities to conclusions of statistical deductions and rationalized the restriction by invoking his ideal observer criterion for validity.

In 1883, Peirce defended calling probabilistic or statistical deduction 'deduction' in analogy to strict demonstration because they both "bring information about the uniform or usual course of things to bear upon the solution of special questions; and the probable argument may approximate indefinitely to demonstration as the ratio named in the first premiss approaches to unity or zero" (CP.2.694). But Peirce emphasized some important differences between the two types of deduction including the following:

A cardinal distinction between the two kinds of inference is, that in demonstrative reasoning the conclusion follows from the existence of the objective facts laid down in the premisses; while in probable reasoning these facts in themselves do not even render the conclusion probable, but account has to be taken of various subjective circumstances – of the manner in which the premises have been obtained, of there being no countervailing considerations, *etc.*; in short, good faith and honesty are essential to good logic in probable reasoning (CP.2.696).

In this passage, Peirce is alluding to the thorny and interrelated topics of identifying a suitable method of random selection from a population and avoiding the acquisition of information about items selected which would invalidate the direct inference from the knowledge of probabilities (the so-called problem of the "reference class"). Peirce contributes very little to the solution of these problems; but he reveals sensitivity to them, acknowledges the subjective and epistemic character of the judgement of probability in probabilistic deduction when that judgement indicates "the modality with which this conclusion is drawn and held for true" and, as we shall see, understands how it prevents assigning probabilities to conclusions of inductive arguments as he construes them.

Let us now take a closer look at Peirce's views about induction or what, in his later work, he called 'quantitative induction'.

In quantitative induction, the hypotheses being subjected to test assert that a proportion of p of Ms are Ps. The problem is to estimate the value of p from data about an n-fold random sample with replacement from the Ms.

Although Peirce was addicted to urn models when discussing induction, there is no need to restrict his ideas to such cases. One might apply his approach to reaching conclusions about the unknown probability of a coin landing heads on a toss on the basis of data about n tosses. But even here the aim is not, for Peirce, the prediction of the limit of relative frequency of heads in an existing infinite sequence of tosses or even in an existing very large sequence. Peirce did not assume that such sequences and their limits exist.

According to Peirce, in induction, the agent observes the relative frequency of Ps in the n-fold sample observed and estimates that the proportion of Ps in the total population will be the same to a given degree of approximation. Peirce sometimes wrote as if the conclusion were the claim that the proportion of Ps in the total population is exactly the same as the observed frequency and that, even when this estimate is in error, it is approximately true. However, Peirce's intention appears to have been that the conclusion of the inductive inference is that the value of p is approximately equal to the observed relative frequency r/n. Thus, strictly speaking, the conclusion of an inductive reference is an interval estimate of the value of p.

Given any value p of the unknown proportion of Ms that are Ps, the binomial distribution can be utilized to compute the probability of any relative frequency among the $n+1$ possible for an n-fold sample. Hence, we can compute the probability of any value of $f(r,n)$ being the interval estimate made if the true value of the unknown proportion of Ms that are Ps is p. Hence, given the value of p, the probability that the estimating function yields an estimate including the value of p is determined. This can be done for the particular estimating function and for every possible value of p. The smallest value k for getting a correct estimate can thereby be identified and the conclusion may be drawn that using the function $f(r,n)$ to make estimates will yield correct estimates with a probability at least equal to k regardless of the true value of p.

The estimating function proposed by Peirce (for the binomial case which is the only one he explicitly considered) was determined by considering some value k greater than 0.5 and less than 1 and calculating, for each value of p, the shortest interval centred around that value such that the probability of the relative frequency of Ps in the n-fold sample falling in that range is at least as great as k and as close to k as possible.

The function $f(r,n,k)$ where k is an additional parameter takes as values the sets (they will be intervals) of values of p such that r/n falls within the interval associated with p with probability k when p is the true value. A careful reading of 'The probability of induction' of 1878 (especially CP.2.685–9) and 'The theory of probable inference' of 1883 (especially 2.698–703 and 2.723–4) should make it unequivocally clear that, at least in those essays, Peirce thought of inductive inference as interval estimation according to rules represented by functions of the type $f(r,n,k)$ just described.

Peirce introduced enough of the algebra required to determine the properties of estimating functions of the type just described to make it clear that as k increased to 1, the interval estimate becomes weaker (*i.e.* the interval estimated becomes longer), as n is increased the interval estimate becomes stronger and, for fixed k and n, if the true value of p were near 0 or near 1, the interval estimate could become extremely definite.

Peirce was often disposed to speak of the probability of the induction; but he quite explicitly denied that the probability attributed to an induction was a statistical probability or a credal probability.

An inductive inference is an inference according to a rule represented by the interval valued function $f(r,n,k)$. The agent knows that on an arbitrary application of this rule (involving the n-fold random sampling with replacement from the population M, the observation of the relative frequency r/n of Ps among the Ms and the interval estimate), the probability of obtaining a correct interval estimate is at least as great as k. Were the agent (counter to fact according to Peirce) to repeat the process *ad infinitum* (sampling from the same population) he would be right with a relative frequency which would converge on k in the long run. (Notice that the procedure does not establish that the investigator will find the correct value in the long run.)

It is tempting to invoke the principles regulating statistical deduction and conclude that, in the specific application of the inductive rule under consideration, the credal probability of a correct estimate is at least equal to k. This would be acceptable prior to actually carrying out the experiment and prior to the agent's finding out the relative frequency of Ps among the Ms sampled.

But once the agent knows the value of r/n, the only basis Peirce could recognize consistently with his own principles for assigning a credal probability to the hypothesis that the estimate is correct is by appeal to a statistical deduction from knowledge of the probability of correct estimation using the rule $f(r,n,k)$ in cases where the value of r/n is known to be equal to the value in the particular case under consideration. That probability is either 0 or 1 depending on the true value of p. If the true value of p could be selected at random from a set of possible values in a manner Peirce jeered at as an absurd ramification of conceptualism, a credal probability for the estimate being correct could be determined *via* statistical deduction:

The relative probability of this or that arrangement of Nature is something which we should have a right to talk about if universes were as plenty as blackberries, if we could put a quantity of them in a bag, shake them well up, draw out a sample and examine them to see what proportion of them had one arrangement and what proportion another (CP.2.684).

Nonetheless, Peirce did allow the right to speak of the probability of an inductive inference; but it is the probability (in the statistical sense) of the correctness of estimates based on the rule $f(r,n,k)$ in general without restriction to cases where the value of r/n takes on the definite value actually observed.

Peirce comments on this in 'The probability of induction' of 1878 (from which the previous citation was taken) as follows:

It appears, then, that in one sense we can, and in another we cannot determine the probability of synthetic inference. When I reason in this way:
   Ninety-nine Cretans in a hundred are liars.
   But Epimenides is a Cretan:
   Therefore, Epimenides is a liar;
I know that reasoning similar to that would carry truth 99 times in 100. But when I reason in the opposite direction:
   Minos, Sarpedon, Rhadamanthus, Deuclion, and Epimenides are all the Cretans I can think of,
   But these are all atrocious liars;
   Therefore, pretty much all Cretans must have been liars;
I do not in the least know how often such reasoning would carry me right. On the other hand, what I do know is that some definite proportion of Cretans must have been liars, and that this proportion can be probably approximated to by an induction from five to six instances. Even in the worst case for the probability of such an inference, that in which about half the Cretans are liars, the ratio so obtained would probably not be in error by more than 1/6. So much I know; but, then, in the present case the inference is pretty much all Cretans are liars, and whether there may not be a special improbability in that I do not know (CP.2.689).

Peirce tended to shun the use of probability in connection with induction in his later writings. Thus, in his comment on 'The doctrine of chances' of 1878 written in 1910, Peirce suggested the terms 'verisimilitude' and 'likelihood' as applicable to the conclusions of inductive reasoning (CP. 2.622).

In what sense did Peirce think inductive rules are self correcting?

Recall that in probabilistic or statistical deduction where on sampling n times at random and with replacement from a population with known p, one might make predictions of the relative frequencies of Ps in the sample, repetition of the process will lead to the prediction of the same relative frequency in the new case as in the old. What will fluctuate from prediction to prediction is the truth value of the prediction made. In the long run, the relative frequency of correct predictions will converge on k; but, of course, no one makes so many predictions about the same population.

Peirce remarked that, in making such a prediction, one might make a mistake on one occasion; but if one reapplied the rule to a new *n*-fold sample from the population, with overwhelming probability the rule of inference used would be vindicated. There is, however, no question of correcting the original prediction made for the first sample. It has been corrected already by counting the returns of that sample:

On the other hand, in induction we say that the proportion ρ of the sample being *P*s, probably there is about the same proportion in the whole lot, or at least, if this happens not to be so, then on continuing the drawings the inference will be, not *vindicated* as in the other case, but *modified* so as to become true. The deduction, then, is probable in this sense, that though its conclusion may in a particular case be falsified, yet similar conclusions (with the same ratio ρ) generally prove approximately true; while induction is probable in this sense, that though it may happen to give a false conclusion, yet in most cases in which the same precept of inference was followed, a different and approximately true inference (with the right value of ρ) would be drawn (CP.2.703).

Peirce is not claiming that induction is self correcting in the sense that following an inductive rule will, in the messianic long run, reveal the true value of ρ. His thesis can be put this way: Either the conclusion reached *via* an inductive rule is correct or, if wrong, the revised estimate emerging from a new attempt at estimation based on a different sample will with probability at least equal to k be correct.

Peirce insisted that any specific induction, if legitimately made, presupposes that the *n*-fold sample from the *M*s is random. It also requires that the property *P* of interest be "predesignate" – *i.e.* designated in advance of the experimental outcome (CP.2.735–40).

Thus, Peirce's inductions are inferences according to rules specified in advance of drawing the inferences where the properties of the rules which make the inferences good ones concern the probability of success in using the rules. These are features of the rules which followers of the Neyman–Pearson approach to confidence interval estimation would insist upon.[4]

In estimating the unknown proportion of *P*s among the *M*s, Peirce saw no problem in assuming that there is such a proportion. The substantive assumption relevant to sceptical doubts about induction is that the *n*-fold sample with replacement is random.

---

[4] Peirce's ideas may be usefully compared with the application of the use of confidence limits to the estimation of binomial parameters by Clopper and Pearson in their (1934).

"But synthetic inference is founded upon a classification of facts, not according to their characters, but according to the manner of obtaining them" (CP.2.692).

The problem of induction for Peirce boils down to determining when sampling from a population is random and establishing that at least in a wide variety of cases our "manner of obtaining" facts is random.

To the extent that this manner meets the requirements, different investigators will each obtain approximately the same estimates from different samples with a very high probability. If reality is characterized as the "object of the final opinion to which sufficient investigation would lead" (CP.2.693), then assuming a reality satisfying this characterization implies the availability to investigators of adequate methods of random sampling. Peirce does not prove *a priori* or attempt, like Reichenbach, to vindicate the claim that some methods of sampling are random enough, but, Kant-like, reasons from the assumption that we can have knowledge of reality through inquiry to the claim that some inductions are legitimate.

Space forbids comparing this interpretation of Peirce's views with alternatives existing in literature. It also prohibits a critical assessment of Peirce's outlook under the interpretation I have proposed. Finally, more needs to be said about Peirce's vision of inquiry seeking a true complete story of the world, the relation of this to Peirce's conception of induction and abduction, and to his view of the fallibility of human knowledge.

In this discussion, my aim has been restricted to pointing out that Peirce's views on induction anticipate the confidence interval method of estimation and cognate procedures for testing statistical hypotheses. Peirce is a forerunner of Neyman and Pearson and, in his understanding of the philosophical importance of characterizing how information about finite samples is relevant to supporting or undermining statistical hypotheses, he anticipated Braithwaite. This is not the only link between Braithwaite's work and Peirce's. Both of them, for example, took Bain's views on belief quite seriously; but the relations between their ideas on statistical inference (or quantitative induction) is an interesting one deserving wider recognition than it has received previously.

Columbia University

REFERENCE

Clopper, C. J. and Pearson, E. S. 1934. The use of confidence or fiducial limits illustrated in the case of the binomial. *Biometrika* **26**, 404–13.

8 *The theory of probable inference: Neyman, Peirce and Braithwaite*

IAN HACKING

This paper will show that the Neyman–Pearson theories of testing hypotheses and of confidence interval estimation are sound theories of probable inference. That is worth arguing because Neyman insisted that there is no such thing as inductive inference; there is only inductive behaviour. Moreover his chief critics, whether they be Bayesian or follow in the footsteps of R. A. Fisher, contend that scientific practice needs a theory of inference, and seriously fault Neyman on that score. I contend that this *contretemps* is provoked by a misunderstanding about inference, and that Neyman has, indeed, provided just the fulfilment of the theory of probable inference that was sketched a century ago by Charles Sanders Peirce.

This volume is a good place to argue the point, for R. B. Braithwaite was the first philosopher to incorporate modern statistical practice into his discussions about science, and his book *Scientific Explanation* makes ample use of the Neyman–Pearson approach. Braithwaite was also an enthusiast for Peirce's ideas about probability and induction, although he did not connect that part of his exposition as closely to Neyman as I now do. It is high time to revive Braithwaite's enthusiasm for Neyman–Pearson statistics. A recent careful article in *The Journal of Philosophy* dismisses the confidence interval approach: 'I believe adequate criticism of this theory is already available in the literature' (Seidenfeld 1978: 710). That is a mistake for which I must share responsibility. My *Logic of Statistical Inference* took vigorous issue with Neyman. This essay is a retraction. I now believe that Neyman, Peirce and Braithwaite were on the right lines to follow in the analysis of inductive arguments.

141

The scandal of philosophy.   A little local history will serve to fix ideas. In 1926 Francis Bacon's third centenary was celebrated in the Senate House of Cambridge University. C. D. Broad concluded a notable lecture by hoping 'that when Bacon's next centenary is celebrated the great work which he set going will be completed; and that Inductive Reasoning, which has long been the glory of Science, will have ceased to be the scandal of Philosophy' (Broad 1952: 142). Whether or not induction is the glory of science, induction, in at least one standard sense of that term, ceased to be the *scandal* of philosophy within a decade, although it will remain problematic for a long time. Each of the three probabilistic routes to the problem of induction had been well worked out within the following ten years, with a good deal of the work done a few hundred yards from the Senate House.

The first way to apply probability to induction is, loosely speaking, "logicist". Keynes had published his version in the 1921 *Treatise on Probability*, which had already been in circulation for some years before appearing in print. The ideas of Harold Jeffreys' *Theory of Probability* were evolving. But the most important work was that of R. A. Fisher. In order to apply logicist conceptions one needs new concepts and *a priori* principles to connect evidence and hypotheses. Sufficiency, ancillarity, likelihood, measures of information and the like were all produced or understood around this time. Indeed had one fully grasped the potentiality of Fisher (1922), that great paper on the foundations of statistics, one might not, even in 1926, have used Broad's phrase, 'the scandal of philosophy'.

Broad lectured on 5 October. The published version of a paper read to the Moral Sciences Club on 26 November repeats the phrase with, perhaps, a touch of derision. We now know that paper well; it is 'Truth and probability' (Ramsey 1931: 197 or 1978: 99). Although Ramsey is less of a subjectivist than is commonly made out, he did begin the second great modern application of probability ideas to induction, the personalist approach rediscovered by Bruno de Finetti and elaborated by L. J. Savage.

Finally there are the ideas associated with Jerzy Neyman and E. S. Pearson: confidence intervals and hypothesis testing using size and power. The two men began their collaboration early in 1926 at University College, London. The confidence interval approach was in fact anticipated by the Harvard mathematician Edwin B. Wilson, a life long devotee of C. S. Peirce.[1] Peirce had long before formulated the chief philosophical idea of the Neyman–Pearson work, that probable inference does not assign a probability to individual hypotheses, but rather draws inferences according to a system with some known rate of success. Despite such earlier formulations, the contributions of Neyman and Pearson were of course independent of Wilson and Peirce. Neyman came to make as trenchant a statement about induction as can be found in the work of any philosopher. There is no such thing as inductive inference – but it does not matter for we can still act wisely by choosing the best patterns of inductive behaviour (Neyman 1957). Hume was right to say that induction is a matter of habit rather than of reason for individual conclusions, but wrong not to see that we can assess our habits. In their classic joint paper of 1933 Neyman and Pearson put the claim more cautiously: 'As far as a particular hypothesis is concerned no test based upon the theory of probability can by itself provide valuable evidence of the truth or falsehood of that hypothesis' (Neyman & Pearson 1933: 290).

It may seem remarkable that Braithwaite made so much use of the ideas of Neyman and Pearson. He was taught about probability and induction by Keynes, and he was Ramsey's contemporary and posthumous editor. Why should he have become so attached to the third great idea about probability and induction, rather than the other two? Perhaps it was partly because there was at the time much enthusiasm, in Cambridge, for some strands in pragmatist thought.

[1] For Wilson's biography and bibliography, see (Hunsaker & Maclane 1973). One of Wilson's teachers was B. O. Peirce and he was fascinated by the entire family, even to the extent of using their genealogical table as a basis for statistical analysis (Wilson & Doering 1926). He recomputed Peirce's experimental studies on the normal distribution (Wilson & Hilferty 1929); the paper has interesting consequences. He described Peirce as 'an expert in making refined physical observations and in reducing them, and a great logician and philosopher' (Wilson 1926a); his admiration for Peirce's writings on probable inference is stated in (Wilson 1926b); the 'confidence interval' paper (Wilson 1927) echoes Peirce in the very title, 'probable inference'. When Neyman graciously attributed the confidence theory to Wilson (Neyman 1952: 222) Wilson said he could not claim that priority. He had merely tried to correct the "logic" of reasoning that employs standard deviation (Wilson 1964: 293). Wilson was rather conservative in demeanour, and dubious of all generalizations not only about statistics but about scientific methodology in general. He did not think of confidence intervals as a universal tool but as a device for a particular job. He expressed at least one reservation about the philosophical basis of Neyman and Pearson (Wilson 1942: 90).

Peirce understood the foundation of the confidence idea. The purpose of this extended footnote is not to claim priority for Wilson but to show that Peirce's ideas are connected, by direct lines of filiation, with a more modern exponent of the technique: his work did not lie entirely fallow. Of course the general confidence idea occurs, in some form, in numerous nineteenth-century writers; A. A. Cournot is often cited in this connection.

'What follows', Ramsey put in a footnote to the concluding section of his paper, 'is almost entirely based on the writings of C. S. Peirce.' Ramsey called his subjective theory 'the logic of consistency'; the subsequent "logic of truth" based on Peircean ideas is less well known today. Much of Braithwaite's exposition runs parallel to Ramsey and Peirce. Braithwaite draws on Peirce most evidently in his "predictionist justification of induction", but I shall show how pertinent it is to the theories of Neyman and Pearson too.

Logic.   What I call a 'logicist' approach to induction holds (a) that correct induction is to be analysed in terms of relations which exist between a body of evidence and an individual hypothesis that it supports or disconfirms, and (b) that these relations are logical in character, so that propositions asserting that these relations hold are *a priori* true or false. Keynes thought that the canonical forms of such propositions are:

(1)   The probability of H on A is p,

and

(2) H is more probable on A than H^* on A^*.

He apparently believed that the probability relation in (1) is like a partial (logical) implication, so that (1) bears an analogy with,

(3) A implies H.

Just as the propositional calculus and *Principia Mathematica* provide constraints on and interconnections among statements of the form (3), so a probability calculus would do the same for (1) and (2). Quantitative probabilities obey the normal rules for probability, while Keynes sketched out a calculus for comparative probabilities, completed in the axioms for "intuitive probability" of Koopman (1940). Koopman used the term 'intuitive' because Keynes held that many judgements of forms (1) and (2) are in the end answerable only to intuitions, on analogy with what G. E. Moore had written about moral judgements in *Principia Ethica*.

Ramsey had a great many criticisms of Keynes' logicist approach, but the most immediately telling "is the obvious one that *there really do not seem to be any such things as the probability relations he describes*".

It may seem surprising to call R. A. Fisher a logicist, but in at least one respect he is close to Keynes. He differs in that he never succumbed to what I call 'the hegemony of probability'. Unlike Keynes he does not believe that all relations between hypotheses and evidence are to be couched in terms of a quantitative, comparative or qualitative relation to probability. Instead he devised his likelihood, significance levels, analysis of contingency tables and so forth. Only when a body of data has a peculiar (and usually deliberately contrived) structure can we engage in that form of "exact inference" that involves overt probability statements. But although Fisher never attached that exaggerated importance to probability statements (1) and (2) that is so common among statisticians and philosophers, he does satisfy my definition of 'logicist' with which I began this section. There are, he held, a variety of *a priori* relations that hold between bodies of data and individual hypotheses, in virtue of which we draw inductive conclusions. My own *Logic of Statistical Inference* unquestioningly made the same assumption. I can now say, at least for the purposes of this paper, what Ramsey said about Keynes: there do not seem to be any such relations. I do not mean that there is no role for likelihood or significance levels *etc.*, but only that these fundamental concepts should not be understood in a logicist way. Nor does my present opinion impugn Fisher's achievements, for it only calls in question a particular way of understanding what he was doing.

It may be tempting to sum up this opinion in the words, 'There is no such thing as a logic of statistical inference'. But to say that is to grant too much to the logicist, for it is to suppose that (1), (2) and the like are the province of inductive logic. On the contrary they are founded on a false analogy with deductive logic. In the next two sections I shall try to destroy the analogy. Unfortunately this involves writing the sort of prose that I do not care to read – with lots of numbered sentences and fine distinctions. For the reader with a similar lack of taste for this sort of work, the results are abstracted in the third section to follow, headed *summary*.

Statement and argument.   Peirce took for granted that the topic of logic is argument, that is, the transition from some sentences to others as in the simplest form,

(4) S, so H.

Here S is a set of sentences (whose conjunction, if finite, we may represent by A). Keynes thought that the topic of inductive logic would focus on probability relations as exemplified by (1) or the comparative (2) or the qualitative (5):

(5)   Given A, H is probable.

I believe that this inclination of Keynes and others was partly fostered by a number of linguistic slides. First, it may be supposed that if (4) is the topic of deductive logic, then inductive logic must concern itself with arguments of the form,

(6)   S, so probably H.

In the following section I shall argue that inductive and deductive logic are both concerned with arguments of the form (4), and that (6) is not a new form of argument but a way of stating the argument of form (4). Postponing that discussion, suppose that as inductive logicians we are concerned with (6). Moreover, we propose to do inductive logic on analogy with deductive logic. We should first consider, then, what the deductive logician does with (4).

Peirce and the older generations of deductive logicians may have thought that (4) should be studied directly, but by the time of *Principia Mathematica* (PM) a false analogy with mathematics had made it standard to regard logic as an axiomatic science of truths, not as a science of rules of inferences, of transitions between sentences. Indeed PM somewhat surprisingly tries to smuggle in its rules of inference among the axioms as one more "primitive proposition". Only in the 1930s did the logician Gerhardt Gentzen restore the older tradition, showing how to get first-order logic from rules of inference alone. Certainly when Keynes wrote he could accept that logic was to be studied not directly as in arguments (4) but by statements of implication such as (3):

(3)   A implies H.

Moreover, as Quine has observed, use and mention were confused. Conditionals of the form 'If A then H' are treated in the propositional calculus by the truth-functional conditional,

(7) $A \supset H$.

The horseshoe was read, 'A materially implies H.' But a statement such as (3) asserts that a relation holds between two sentences, namely the relation of implication. It mentions (to use Quine's jargon) these sentences A and H, and says something about them. But the conditional (7) uses these sentences to form a complex sentence. 'A implies H' and 'If A, then H' are *not* synonymous. Now the argument 'A, so H' is truth-functionally valid if and only if (7) is a tautology, or a theorem of the propositional calculus. So it was thought that the argument 'A, so probably H' must be inductively valid if and only if a statement such as (5), 'Given A, H is probable' is a theorem of a probability calculus – or if not (5) then some quantitative statement such as (1). Moreover, it was thought that (1) or (5) would be parts of a probability calculus in the way that (7) is, *and, at the same time*, that (1) and (5) express relations between the propositions A and H. That is the result of the confusion of use and mention that conflates (3) and (7). On the relational view of probability, (1) is like (3), expressing a relation between propositions, but then it is not like (7), *i.e.* not like an ingredient in the propositional calculus. But, like most confusions of use and mention, this error would have been unlikely to do any harm in itself, had it not been compounded with a more serious mistake.

Probable inference.   An argument of the form,
  (4) S, so H,
will be valid if and only if the statement of the form,
  (3) A implies H,
is true. So can we not say that an argument of the form,
  (6) S, so probably H,
is inductively valid if and only if the relational statement of the form,
  (5)   Given A, H is probable,
is true? It may only be a matter of taste to prefer Gentzen's regarding logic as a matter of argument (such as (4)) as opposed to *Principia Mathematica*'s treating it as a matter of axiomatizing sentences (such as (3)). May we not then as a matter of taste go along with Keynes and make (1) or (5) the topic of inductive logic? I think not, because I believe that the 'probably' in (6) is like the 'necessarily' in one reading of,
  (7) S, so necessarily H.
Now one can say 'necessarily H' to mean that H is a logically necessary proposition. But in the context of (7) the common use of 'necessarily' is to indicate that H follows necessarily from S. That is, 'necessarily' modifies the argument, the way in which H is reached. It is not an adjective modifying H *simpliciter*.

  Thus (7) is not a different argument from (4), 'S, so H.' It is a different way of expressing the same argument, at the same time indicating the force with which one may draw the conclusion from the premisses. Likewise (6) is not a different argument from (4); the 'probably' in (6) modifies the argument, not the conclusion H.

  Yet it will be protested that we can say simply,
  (8)   Probably H.

Why cannot (8) be the conclusion of the argument (6)? In answer I shall recall two observations from the time when Oxford linguistic philosophy was at its apogee. One is due to D. G. Brown, and concerns the word 'inference'; the other is by Stephen Toulmin, about the word 'probably'.

I retain the logician's use of the word 'argument': a string of sentences with a concluding sentence. Brown (1955) says that 'inference' does not mean 'argument' in this sense. 'S, so H' represents an argument, not an inference. The *inference* is H. An inference is a concluding statement made in an argumentative context. Incidentally, Peirce sometimes appears to be using the word in this way, as do most writers of his time who treat of logic. But unlike Brown I do not claim that this is the only or even the favoured way of using the word 'inference', only that it is a correct and for us instructive way to use it.

Next examine what Toulmin (1965) says about 'probably' and combine it with Brown's idea about inference. He says that 'probably H' is not a statement about H. It is a way of stating H. 'Probably' is as it were in parentheses, employed to make the statement H in a certain way. To state H in normal conversation is to imply that H is warranted, or that one will stick by H, backing it up with reasons, and so forth. To say 'probably H' is to state H in a guarded way. That is, to imply that one's back-up for H is less than conclusive, and that one grants that H might turn out to be false, although one doesn't expect that.

Now we can combine Brown and Toulmin to comment on the form of words,

(6) S, so probably H.

This is just the argument, 'S, so H'. In either case the inference is H. This inference is a statement which may be guarded, like any other, by prefixing 'probably'. But in the context of the argument one knows the source of the caution that prompts this guarded statement: the argument is not a conclusive one.

It will be noticed that an argument may be inconclusive for more than one reason. The premises may not entail the conclusion, so the argument may be offered as a merely probable argument. But, also, the premises may be stated in a guarded way, as when one says, 'probably S, so probably H'. If S entailed H I would not call that a probable argument. But the function of 'probably' in the conclusion has the same core role as in probable argument, namely to express caution about the statement H. Naturally none of this takes us any positive distance towards understanding probable argument. These have been merely negative remarks, intended to dispel the illusion that inductive logic should concentrate on statements of the form,

(1)   The probability of H on A is p.

Perhaps I should add that Ramsey seems to have subscribed to the view that "probable inference" would be a theory of drawing inferences to conclusions (1). 'The suggestion of Mr Keynes that [Hume] can be got round by regarding induction as a form of probable inference cannot in my view be maintained. But to suppose that the situation which results from this is a scandal to philosophy is, I think, a mistake.' Ramsey wrote these words in the course of elaborating on Peirce. I differ from Ramsey chiefly because I believe that what results from Peirce's ideas is exactly what should be called a theory of probable inference, and that only logicism prevented one from seeing that. Peirce himself always called it either probable inference or probable argument. E. B. Wilson, one suspects with Peirce in mind, and writing at exactly the same time as Ramsey, titled his version of confidence theory just that: 'probable inference'.

Summary of the preceding two sections.   Logic is concerned with the transition between sentences, *i.e.* with arguments 'S, so H'. In such argument the inference is the conclusion H. An inference is not a transition but a statement. A statement may be expressed in a guarded way, *e.g.* by prefixing the word 'probably'. Thus the argument 'S, so probably H' is not an argument to a conclusion 'probably H' but an argument to the conclusion H, which inference is expressed in a guarded way. Inductive logic is not the logic of statements of the form, 'the probability of H on A is p'. These have been taken to be the province of inductive logic for a mixture of poor reasons. Such statements were wrongly taken to be the conclusion of any quantitative inductive argument: they are not. Moreover certain false analogies with implication, and confusions of use and mention, abet this tendency among earlier inductive logicians.

A genus of arguments.   A valid demonstrative argument, wrote Peirce, 'is a member of a genus of arguments all constructed in the same way, and such that, when their premises are real facts, their conclusions are so also' (Peirce 2.649). [2] Today's logician would give a semantic account of this in terms of logical consequence. That requires some apparatus. A model structure for a class of sentences closed under suitable syntactic operations consists of a class of models or interpretations of the sentences, together with functions that have the effect of assigning truth values to sentences in the models. H is a logical consequence of S when H is true in all models in which every member of S is true.

Evidently there are formal devices that will define a notion of partial logical consequence, even one which is amenable to a probability measure. That might seem the right way to explicate Peirce's frequent pronouncements to the effect that 'We may, therefore, define the probability of a mode of arguments as the proportion of cases in which it carries truth with it' (2.650). That, however, would not do. He had more than proportion in mind, for in the same paragraph he writes: 'In the long run there is a real fact which corresponds to the idea of probability, and it is that a given mode of inference sometimes proves successful and sometimes not, and that in a ratio ultimately fixed.' A model theoretic approach needs to be augmented by this "real fact" of stable long run frequency.

Frequency. When Peirce was young he was a nominalist who had no truck with contrary-to-fact conditionals. The probability of getting an ace in tossing a die is a relative frequency of aces in a sequence actually performed. A diamond at the bottom of the sea is hard only if actually tested. As Peirce matured he adopted a scholastic realism and came to see that what would be the case is an integral part of the way the world is. The diamond is hard if it would have proved hard; the probability of getting an ace is the "would-be" of the die, as Peirce says in what Braithwaite rightly calls a 'striking description' (Peirce 2.664; Braithwaite 1953: 187).

Braithwaite's account of objective probability has not been significantly bettered. Probability in this sense does not mean "relative frequency", but probabilities are typically manifested by stable frequencies. Probabilities conform to the usual probability axioms which have among their consequences the essential connection between individual and repeated trials, the weak law of large num-

---

[2] References to Peirce are in the customary form of volume and paragraph numbers of his collected papers (1932).

bers proved by Bernoulli. Probabilities are to be thought of as theoretical properties, with a certain looseness of fit to the observed world. Part of this fit is judged by rules for testing statistical hypotheses along the lines described by Neyman and Pearson. It is a "frequency view of probability" in which probability is a dispositional property, subject "to the whole gamut of logical and epistemological considerations developed in" Chapter VI of *Scientific Explanation*. Naturally various niceties will be added to this account, *e.g.* Patrick Suppes' programme of proving representation theorems for propensities, and there will be debates about what the propensities are propensities *of*, whether truth about probability requires indeterminism, and the relation between single cases and the long run (as if Bernoulli had not settled that). But Braithwaite's account is a good careful philosophical explication of what statisticians mean, as when Fisher talks about relative frequencies in hypothetical infinite populations. Neyman thanked Braithwaite for this quite explicitly and it is entirely in the spirit of Peirce. How are we to apply it to Peirce's conception of probable inference?

The truth-producing virtue.   Peirce defined *validity* as 'the possession by an argumentation or inference of that sort of efficiency in leading to the truth, which it professes to have' (2.779). He went on to call this the 'truth-producing virtue'. In the case of inductive argument, this virtue consists in the fact that the argument is of a form which would lead to the truth, "for the most part".

It may be conceived, and often is conceived, that induction lends a probability to its conclusion. Now that is not the way in which induction leads to the truth. It lends no definite probability to its conclusion. It is nonsense to talk of the probability of a law, as if we could pick universes out of a grab-bag and find in what proportion of them the law held good (2.780).

This is equally the opinion of Wilson, Neyman and Pearson. Peirce did not adequately develop the mathematical implications of his insight, so we pass at once to more modern procedures.

To take the simplest but sufficiently general case, suppose we require an estimate of a quantity such as the chance of getting an ace with a die or the mean of a normally distributed population. We incorporate idealized statistical assumptions about the chance set-up into a statistical model of a family of distributions with unknown parameter θ (*e.g.* the chance of an ace or the mean height of the population). A class of possible observations is represented in a sample space.

An interval estimator for θ is a function f with the following properties. First, it is a function from points in the same space to intervals on the parameter space. Secondly, for any θ in the parameter space the probability on θ of getting a sample x that covers θ is (say) 95 per cent.

Hence, assuming that the statistical model is correct, the probability – regardless of the value of θ – of making a correct estimate using f is 95 per cent. Let us then make an observation, say x_0. Using f we estimate that the unknown true value of θ, call it θ_0, is in $f(x_0)$.

This estimate is reached by a procedure that gives correct results 95 per cent of the time. All authors in this tradition rightly insist that we cannot therefore attach 95 per cent probability to the statement 'θ_0 is in $f(x_0)$.' Yet, contrary to this tradition, we may still talk about inference. There are two premisses, one stating the assumptions built into the statistical model, and the other reporting the observation. This leads to the following argument:

(i)   (A premiss setting forth the statistical model.)
(ii)   The sample x_0 was observed.

So,   (iii) θ_0 is in $f(x_0)$.

The statement (iii) is the inference. One may express the inference in a guarded way: 'Probably θ_0 is in $f(x_0)$.' That does not mean that (iii) has some definite probability, only that one is making a cautious statement. The most informative way to express reservations about (iii) is to say that it is reached by a method that has a 95 per cent probability of being correct. An exactly parallel account may be given for inferences made using a fixed significance level or size of test.

Critics of Neyman and Pearson sometimes say that confidence intervals perpetrate a confidence trick on the innocent research worker. The routine technician conducts an experiment and obtains a 95 per cent confidence interval. But even when this person has been taught somewhere along the line that you cannot attach a 95 per cent probability to the statement, 'θ_0 is in $f(x_0)$', what the interval *means* to the researcher is just, 'the probability that θ_0 is in $f(x_0)$ is 95 per cent'. That is what the confidence interval "feels like" to the research worker, and that is how it is often used, or so the critics say. I think it is only the logicist instincts of the critics, and their false view of language, that leads them to impugn the research worker. The person in the laboratory does draw an inference (Neyman was wrong about that) but the inference is just, 'θ_0 is in $f(x_0)$.' That is the inference one wants, together with an indication of the reliability of the inference. Indeed the older version of the same way of inferring is still taught to physics students; one concludes an experimental measurement with an indication of the probable error. The old term 'probable error' is a horror, but the idea is sound enough.

Efficiency. Peirce, as quoted above, said that the validity of an argument is a kind of efficiency, but only the precise analysis of Neyman and Pearson makes clear the aptness of this metaphor. Their 1933 paper employs the word in its title – 'the most efficient tests of statistical hypotheses'. In our terms the problem of efficiency arises because if there are any 95 per cent estimators there are typically too many. How to choose the best 95 per cent estimator, or the best 95 per cent test?

Peirce took a thoroughly practical attitude towards reasoning. Good reasoning serves certain ends. He idealized the end as the pursuit of truth. But as Isaac Levi says in a related connection, this pursuit involves two goals, a desire to be correct, and "relief from agnosticism" (Levi 1967: 62). One can minimize mistakes by never engaging in ampliative inference; it is the desire for more information that leads to risk. 'Believe truth! Shun error! – These we see are two materially different laws', as William James puts a not unrelated point with characteristic flair.[3] In terms of estimators one can serve both these masters, truth and informativeness, if one can find a 95 per cent estimator which, for all θ, gives uniformly narrowest estimates. That is, we would prefer an estimator which, at given security level, produces narrower estimates than other estimators, regardless of the sample observed and the true known value of the parameter. Unfortunately this tidy solution is mathematically available only in the simplest cases.

E. S. Pearson has described how he and Neyman tried to formulate this kind of problem in general terms and arrived at some of the solutions that have since become standard (Pearson 1966). It has been increasingly realized that unique solutions become available

[3] (James 1897: 18). John Etchemendy drew my attention to the parallel between William James' injunction and the Neyman–Pearson motto, 'control the size of a test; maximize its power'.

only when one specifies more precisely what the estimate or test is for. Mathematical work in this branch of statistics has been transformed into decision theory. I do not take it to be a criticism of the Neyman–Pearson theory that there is no mindless procedure for applying it to any statistical problem. It is surely a virtue sometimes to be speechless or to invite more data or to ask exactly what is the point of the inquiry. A great many real-life problems, even after they have been forced into parametric models, simply do not have uniquely best solutions without further analysis, abstraction or experimentation.

My remarks are not intended to endorse every Neyman–Pearson style solution on the books. The most serious criticism of the method is that it does not always use the information which may, perhaps by ill luck, be cast on the experimenter's plate. A policy which in general is a good one might on a particular occasion reveal itself as inept, as in the distinction between before-trial and after-trial betting (Hacking 1965: 95–102). But it is important to recall that Neyman–Pearson techniques, understood as rules of inference, are not rules that tell you what you must do on every occasion. To see this, some further discussion of the use of such rules is in order.

Acceptance.   The theory of testing hypotheses was first presented as a decision procedure with two outcomes: either accept or reject the hypothesis under test. Braithwaite observed that 'don't reject' does not imply 'accept'. He himself had something of a falsificationist methodology, according to which only falsifiable statements have real content; since he was examining the meaning of statistical hypotheses, he emphasized rejection. The terminology of 'accept' and 'reject' had won some favour among statisticians because of applications to quality control where one is literally accepting a consignment of light bulbs. It was soon realized that testing readily generalized to three or more options, *e.g.* 'accept', 'reject', 'suspend judgement until examining a further batch'. The idea of accepting hypotheses did no harm in statistics. The same may not be said of philosophy.

The philosopher's exemplar of a rule of inference is *modus ponens*, 'From A and $A \supset B$ to infer B.' It seems absurd to assert the premisses while denying the conclusion, so such rules have been compared to obligations: they *must* be followed, on pain of unintelligibility. Traditionally one said that if the premisses are true, the conclusion must be true, or follows necessarily. There is little harm in comparing rules of inference in deductive logic to commands.

Since in an ampliative argument the conclusion does not necessarily follow, we should not think of inductive rules as commands; perhaps they are more like licences, granting permission. Although such deontic comparisons are largely futile, they can prevent some obfuscation. Consider the lottery paradox. I learn that a fair lottery has 10 000 tickets, of which only one will receive a prize. I am offered ticket 401. Reflecting on the information to hand, I infer, *i.e.* state, with hardly any reservations, that ticket 401 will not win. I have used a form of argument whose probability is 99.99 per cent; only one time in 10 000 would I err if I regularly reasoned in this way. Then it is protested: in parity I must draw the same inference about tickets 402, 403, 9999, 1, 2, . . ., 399. So I must infer that no ticket will win!

I "must" infer nothing of the sort. I need infer nothing whatsoever. To infer is to make a statement on the basis of reasoning. When offered 401 I infer with almost no hesitation that it will not win. If I am offered my pick of any of the tickets, I infer with equal confidence that whatever ticket I pick, it will not win. But I do not propose to make 10 000 inferences; I should get hoarse. But if I make several inferences taken together, I may have more reservations about the joint inference than about any one inference made individually. If I am offered all the tickets 401–500, I shall say that *probably* none will win. That inference is drawn according to a rule with 99 per cent probability, and perhaps that is when I start expressing caution.

Henry Kyburg, who has placed much store by the lottery paradox, agrees one can infer that ticket 401 won't win. But he goes on to propose that I should "admit into my corpus of belief" – which is how he understands inferring – all the propositions of the form, 'ticket *n* won't win'. According to him, I infer each of '*n* won't win', from $n = 1$ to $n = 9999$. But he denies that if I infer A and infer B, then I must infer $A \& B$ (Kyburg 1970). On my account, to infer is to perform a certain act, to wit, making a statement. To state A and to state B, on the same occasion that one declines to state $A\&B$, can only invite an inquiry as to what on earth one means. But to state A guardedly on the same occasion as one states B guardedly is consistent with expressing even greater caution about the joint truth of A and B. Perhaps I am taking a position, on the lottery paradox, that resembles Kyburg's. But it seems much less tendentious when couched in terms of guarded inference. The school of Neyman and Pearson sometimes said of course it dealt with inference, for it provided a theory of acts, and saying that θ_0 is in $f(x_0)$ is just as much an act as buying θ_0 board feet of lumber. Although this is sometimes dismissed as idle sophism, it is not. The theory of inference must be part of what J. L. Austin came to call the theory of speech acts.

The lottery paradox is analogous to the situation with statistical inference because Peirce, Neyman and Braithwaite transformed inverse inference – in which one infers a statistical hypothesis from experimental observations – into direct inference, in which one infers a particular case from knowledge of a statistical distribution. Knowing the distribution of the estimator f, and having observed x_0, one infers that θ_0 is in $f(x_0)$. But since there may be any number of 95 per cent estimators, one could, if one drew all the inferences that they license, infer any number of incompatible statements. The lottery paradox is exacerbated. But not only is one not obliged to infer anything that follows by any 95 per cent rule, but also there is, on Peirce's methodology, a fundamental reason for using just one rule. This arises from an aspect of his theory that I have thus far left aside: habit.

Habit. Hume notoriously concluded that inductive conclusions cannot be justified by reason, and are drawn only as a matter of habit. Peirce made a few direct allusions to Hume, but he did adopt many doctrines from the Scottish school of philosophers, Reid, Steward, Bain and so forth. The Scots dominated mid nineteenth-century philosophy and it was natural for him to take up their idea that belief is to be understood in terms of habits of action. Reasoning too was a matter of forming habits, the habits of drawing conclusions. But whereas Hume had thought habit and custom are not susceptible of criticism, "habit", for Peirce, had no connotations of irrationality. One seeks habits that will tend to lead to the truth. Use of a 95 per cent estimator is one such habit. But use of an arbitrary 95 per cent estimator, or of a lot of conflicting estimators, is not such a habit. The rational agent will choose an estimator with some optimal characteristics. But it is not required for the theory to say, in general, what is optimal. That will depend, perhaps, on the purposes to hand. Note, moreover, that the present analysis does not commit one to each of the particular solutions which have been urged by the Neyman–Pearson school. I am contending only that the theory provides an account of inference.

Yet there remains a question. What is the merit of adopting a habit with the truth-producing virtue, if one is in fact to deploy the habit only once? If I were to reason many times by the use of a habit, then I might be rewarded by being right most of the time, but what about those cases when only one inference is to be drawn? How can long run virtues justify short run policies? My *Logic of Statistical Inference* will not be accused of understating this difficulty, yet in effect it merely repeats Peirce (2.652). How can Peirce have persisted in his theory of inference when he was so aware of this objection? One answer is, in his words, akin to Faith, Hope, and Charity (2.655). The agent that does not identify his interests with those of all mankind is irrational. I may deploy my habit only once, but my act of reasoning is only one among a host of human acts. Now this is a "nominalist" answer, attempting to justify a habit in terms of a long run frequency among actual choices, albeit not my choices. But it is defective. We believe that an argument with the truth producing virtue would be a good argument even if it were the last argument ever to be propounded; even if our race were about to become extinct. Perhaps 'Hope' is supposed to rule that out. But there is another difficulty to which Hope does not cater. We may devise a particular Neyman–Pearson solution to a problem of a sort which we have no reason to hope will ever occur again, simply because the statistical model applies to a bizarre and fortuitous concatenation of circumstances. We may have every reason to hope that our inference is unique. How then do the long run merits of a habit justify the inference?

The nominalist answer of Faith, Hope and Charity is unsatisfactory. But as Peirce insisted, a passage from nominalism to what he called scholastic realism is part of the maturing of the philosopher. We have to admit what "would be" into our account of what the universe is. What would happen in this world is an irreducible fact about this world. More surprisingly, benefits which "would accrue if" must be included among our roster of what counts as reason.

Induction.   According to Peirce there are two kinds of ampliative reasoning. One he called 'probable argument', and the other he variously called 'hypothesis' or 'abduction'. He kept on changing his opinion about the relative domain of each. He long hoped to get some kind of probability for abductive inferences, 'but when I finally succeeded in clearing the matter up, the fact shone out that probability proper had nothing to do with the validity of Abduction' (2.102). Now the method of hypothesis or abduction is that of finding explanations for otherwise inexplicable collections of facts. A great deal of what Peirce says about hypothesis fits Braithwaite's classic account of the hypothetico–deductive method. Peirce thought that induction is for testing hypotheses that have been conjectured as explanations, and he foresaw the enormous role that statistical hypotheses were to have in twentieth-century science. That branch of logic which shows how to test them is to be called 'quantitative induction' (2.758). But it relates 'to but a small part of the Logic of Scientific Investigation' (2.751). I began by quoting Broad: 'induction, which is the glory of Science'. As Peirce and I understand the term, induction is *not* the glory of Science. Induction is a matter of good habits. In different epochs we attend to different glories of Science. Ancient Greece taught that the glory lies in the power to produce demonstrative conclusions, and took geometry to be the model for Science. In our day we much emphasize the power to break the old habits of thought and create new, albeit fallible, versions of the Universe. It is not a defect in a theory of induction that it should fail to account for all or even the most important parts of the logic of scientific investigation.

It is of course possible to use the word 'induction' to mean whatever it is that science does, and to cover every species of ampliative reasoning. Perhaps that is a Baconian use of the word, one found in Whewell's histories and philosophy of 'The Inductive Sciences'. It is wiser to follow Peirce, and allow of at least two kinds of ampliative reasoning. Then induction becomes a matter of humdrum habit. A theory of induction should indeed be relevant to Hume, who notoriously cared little about most of the things we now call the glories of Science. Instead he worried about everyday reasoning – not reasoning that would shift our concepts and redeploy our imaginations, but the plain, the humdrum, the habitual.

In answering Hume we need not tell him the basis for every belief. On the contrary, we need only tell him how to infer, on the basis of some propositions to which he already subscribes, some new statements about the future or about the world in general. That is precisely what is done by the Neyman–Pearson theory of probable inference. Common people no more use the categories of that theory than they employ first-order deductive logic. But just as the latter is the idealization that furnishes our best present understanding of demonstrative reasoning, so the former is one way to abstract and analyse the inchoate structure of inductive argument. But there may be different abstractions of the same thing, both instructive, yet not isomorphic. That is the present state of our understanding of inductive argument. The confidence approach furnishes one sort of model of learning from experience. The neo–Bayesian subjective tack furnishes another. I have never seen any incompatibility between the two. It is seldom noticed that F. P. Ramsey's 'Truth and probability' (Ramsey 1978) ends by trying to combine his subjective theory – 'the logic of consistency' – with Peirce's approach – 'the logic of truth'. He goes so far as to suggest that only such an embedding will provide the real explanation of why degrees of belief satisfy the probability axioms.

Stanford University

REFERENCES

Braithwaite, R. B. 1953. *Scientific Explanation*. Cambridge.

Broad, C. D. 1952. *Ethics and the History of Philosophy*. London.

Brown, D. G. 1955. The nature of inference, *The Philosophical Review* **64**, 351–69.

Fisher, R. A. 1922. On the mathematical foundations of theoretical statistics. *Philosophical Transactions of the Royal Society of London A* **222**, 309–68.

Hacking, Ian. 1965. *Logic of Statistical Inference*. Cambridge.

Hunsaker, Jerome and Maclane, Saunders. 1973. Edwin Bidwell Wilson, *Biographical Memoirs of the National Academy of Sciences*. **43**, 285–320.

James, William. 1897. *The Will to Believe*. London.

Koopman, B. O. 1940. The axioms and algebra of intuitive probability, *Annals of Mathematics* **41**, 269–92.

Kyburg, Henry E., Jr 1970. Conjunctivitis. *Induction, Acceptance and Rational Belief*, ed. Marshall Swain, pp. 55–82. Dordrecht.

Levi, Isaac. 1967. *Gambling with Truth*. London.

Neyman, Jerzy. 1952. *Lectures and Conferences on Mathematical Statistics and Probability*, Second edition. Washington.

Neyman, Jerzy. 1957. 'Inductive behavior' as a basic concept in the philosophy of science, *Revue de l'Institute Internationale de Statistique* **25**, 7–22.

Neyman, Jerzy, and Pearson, E. S. 1933. On the problem of the most efficient tests of statistical hypotheses, *Philosophical Transactions of the Royal Society of London A* **231**, 289–337.

Pearson, E. S. 1966. The Neyman–Pearson story 1926–34. In *Research Papers in Statistics*, ed. F. N. David, pp. 1–24. New York.

Peirce, C. S. 1932. *The Collected Papers of Charles Sanders Peirce*, eds. C. Hartshorne and P. Weiss. Cambridge, Mass.

Ramsey, F. P. 1931. *The Foundations of Mathematics and Other Logical Essays*, ed. R. B. Braithwaite. London.

Ramsey, F. P. 1978. *Foundations*, ed. D. H. Mellor. London.

Seidenfeld, Teddy. 1978. Direct inference and inverse inference. *The Journal of Philosophy* **75**, 709–30.

Toulmin, S. E. 1950. 'Probability'. I, *Aristotelian Society Supplementary Volume*, **24**, 27–62.

Wilson, Edwin Bidwell. 1926a. Statistical inference. *Science* **43**, 289–96.

Wilson, Edwin Bidwell. 1926b. Empiricism and rationalism. *Science* **44**, 47–57.

Wilson, Edwin Bidwell. 1927. Probable inference, the law of succession and statistical inference. *The Journal of the American Statistical Association* **22**, 209–12.

Wilson, Edwin Bidwell. 1942. On confidence intervals. *Proceedings of the National Academy of Sciences* **28**, 88–93.

Wilson, Edwin Bidwell. 1964. Comparative experiment and observed association. *Proceedings of the National Academy of Sciences* **51**, 288–93.

Wilson, Edwin Bidwell and Doering, Carl R. 1926. The elder Peirces. *Proceedings of the National Academy of Sciences* **12**, 424–32.

Wilson, Edwin Bidwell and Hilferty, Margaret M. 1929. A note on C. S. Peirce's experimental discussion of the law of errors. *Proceedings of the National Academy of Sciences* **15**, 120–5.

9 Statistical statements: their meaning, acceptance, and use

HENRY E. KYBURG, JR

1 In *Scientific Explanation* (1953) Braithwaite offers a novel and picturesque interpretation of statistical statements. It is offered as a discussion of the meaning of 'probability' as it occurs in scientific hypotheses. In the discussion that follows, I shall use the more neutral term 'measure'. Since "statistical hypotheses" in science more often concern the distribution of a certain random quantity in a class, than merely the measure of some subclass in a reference class, 'measure' seems more appropriate than 'probability'; and this choice of terminology allows us to reserve 'probability' for purposes which will become apparent in due course.

Let us consider the meaning of 'the measure of As among Bs is p'. When B is non-empty and finite, this may be construed simply as the *proportion* of As among Bs (Braithwaite 1953: 122). According to Braithwaite, however, we cannot construe the reference class B as finite in the case of a *scientific* hypothesis. 'For if it were assumed that the class of reference had only a finite number of members, the generalization would be restricted to apply only to a limited number of its instances, and would thus not have the generality which we require of a generalization for it to be ranked as a scientific hypothesis' (Braithwaite 1953: 123).

On the face of the matter, this is not persuasive. A cosmological theory according to which there were only a finite number of massive bodies in the Universe would, on this ground, deprive a theory of gravitation of its status as a "scientific" hypothesis; statements in biology concerning the distribution of colours or weights in a certain species would not be allowed as "scientific", since we have good grounds for supposing that the number of instances of any species is finite; similar remarks would apply to hypotheses in

161

the social and behavioural sciences, where we have good reason to suppose that the actual reference class is finite. On the other hand, one can easily sympathize with the intuition: the distributions derived from quantum mechanics seem somehow much more "scientific" than, say, the hypothesis concerning the measure of the set of voters in a small town who will vote Republican in the next election.

We may distinguish three levels of generality in statistical hypotheses. At the lowest level, we have such hypotheses as: 'Between 51 per cent and 55 per cent of the voters in the town of Lyons will vote Republican in the 1980 election.' We may wish to withhold the honorific 'scientific' from such hypotheses, but they often function in the same way as more "scientific" hypotheses in guiding our decisions, and they are often confirmed in much the same way as more scientific hypotheses. In particular, we usually confirm such hypotheses by examining a sample, and Braithwaite's rule of rejection would seem to apply to them as well as to the more general hypotheses he prefers to discuss. Since the class of reference (the set of voters in Lyons in the 1980 election) is not only bounded, but could in principle be enumerated – modulo minor changes – by going through the list of registered voters, it is clear that this is not the sort of hypothesis that Braithwaite wants to discuss. Let us nevertheless keep this sort of hypothesis in mind, and call it a statistical hypothesis of class I. Here are some other examples: We wish to know the distribution of hat sizes in the senior class at a large university for the sake of laying in a supply of academic caps for graduation; we might use a sample of seniors as the basis for a rough hypothesis as to that distribution. We want to know how many people in a certain city would buy a certain product, in order to decide whether or not to open a new outlet for our product there. We want to know how serious a problem the corn rootworm borer is in a certain county.

In each of these cases what interests us is a certain distribution or measure in a finite, specified population. We could, in principle, survey the entire population, but considerations of practicality and expense lead us instead to support a statistical hypothesis about the finite specified population by surveying a sample of it.

At the next level of generality, we may consider statistical hypotheses in the biological and social sciences. Here we may consider such hypotheses as 'the wingspan of species X is roughly normally distributed with a mean of m and a variance of s^2', or 'the mean lethal dose of substance Z for mice is k milligrams per kilogram of body weight', or 'the annual risk of a coal miner in a mining operation of a certain description is p'. These hypotheses differ from the preceding ones in that the class of reference is open: we don't know how many members of species X there will be; we don't know how many mice there will be; we don't know how long miners will continue to work in mining operations of the sort described. These classes certainly satisfy Braithwaite's requirement that 'the class of reference in a scientific hypothesis must be a class which is not limited in advance by the way in which the expression of the hypothesis is interpreted (1953: 123).

Nevertheless, we know perfectly well that these classes are finite. By a judicious use of large exponents, we can even easily enough name finite integers that exceed the cardinality of these classes. (My favourite is the Googolplex: $(10)^{100^{100}}$.) Now as far as the interpretation of the parameters in statistical hypotheses is concerned, all that interferes with interpreting them as class ratios is the possibility that the class of reference is literally infinite. Thus in all these cases, though we do not know the cardinality of the classes concerned, we may give the *semantics* of statistical hypotheses in terms of class ratios in finite classes. I shall call these hypotheses, hypotheses of class II.

It is less tempting to withhold the honorific adjective 'scientific' from these hypotheses than to withhold it from hypotheses of class I. Nevertheless, hypotheses of these two classes are difficult to distinguish sharply. In the case of the voters of Lyons, we have the registration list. In the case of the cornborers in Wayne county, we have no list; we could make an exhaustive survey of every field, counting the number of damaged cornstalks, instead of taking a sample, but this is less feasible than going through the complete list of registered voters. We could, in principle, wait until the last mine of the sort concerned had closed down, and then enumerate the fatal accidents – but this would not provide us with information relevant to our decisions regarding such mines *now*. We can hardly wait until the species of mice no longer exists, but if the subject of our experiment were snowgeese, there is a reasonable chance that in a number of years there will be no more. Whether we apply or withhold the term 'scientific' from such hypotheses as those of class II therefore seems somewhat arbitrary. In either event, hypotheses of class I as well as those of class II are important as guides to action and decision, as well as to our understanding of the world. They function in much the same way as hypotheses of class III, to which we now turn.

The statistical assertions of quantum mechanics are of course the paradigm instances of scientific statistical hypotheses worthy of the name. We need not go quite so far afield for examples, however. Consider 'The decay rate of radium is given by the exponential distribution with parameter lambda', or 'The linkage coefficient of gene A and gene B in Drosophila is beta.' It might be tempting to construe these as hypotheses of type II: according to the latest popular science, there are only a finite number of radium atoms in the universe; it is certainly reasonable to assume that the total number of Drosophila chromosome recombinations, past, present, and future, is finite. But we would want to say that the parameters lambda and beta were what they were, regardless of how many or how few radium atoms or Drosophila chromosomes there are. The reason is that these parameters play a role in quite general theories.

Let us look at a more down-to-earth example. Consider a die of uniform density being thrown on a smooth horizontal surface. It is not hard to argue that the distribution of spots on the uppermost side in a sequence of throws is independent of the distribution of the components of momentum and locus of the die at the time of release, subject to very mild constraints. If the distribution of these components is reasonably large and reasonably continuous, it will follow from the laws of mechanics that the measure of "ones" will be 1/6, for example. At the same time we know, of course, that the chances are very much against getting exactly 1/6 aces on any finite number of throws of the die. The number 1/6 is nevertheless an important and useful number in describing the behaviour of this ideal die, and *a fortiori* in describing the behaviour of an actual die being thrown on an actual surface, provided the set-up closely approximates our ideal set-up. We often express this by saying that the *chance* of getting an ace on a throw of the die is 1/6. We are thus led to construe the parameters in more serious scientific hypotheses – the beta and lambda of the previous examples – as measuring chances. Although Braithwaite prefers to use the term 'probability', rather than 'chance', it seems quite clear that it is this abstract notion of chance that he is trying to capture in his Briareus model.

Let us leave statistical hypotheses of the first two classes to one side, for the moment, and consider the semantics of chance. It is clear that an extensional semantics – in the most natural sense of the term 'extensional', in which we are concerned with actual frequencies in the actual world – will not do. Even if no one had ever constructed or rolled a die that was an approximation to the ideal one, we would nevertheless want to say that the chance of an ace coming up in a mechanical system of the sort described was 1/6. If in fact there were a lot of such trials on such systems in the world, and none of them yielded an ace, we would be led to deny that the chance was a sixth – but the chain of argument would be indirect: we would take such an actual result as casting severe doubt on laws of mechanics and the assumptions we had made in deriving the value 1/6.

We are therefore constrained to adopt some sort of possible world semantics. This is just what Braithwaite's Briareus model is designed to provide. The statistical hypothesis that the number of aces in a sequence of n throws of a die is binomially distributed with parameters n and 1/6, corresponds to a Briareus model with n urns, each of which contains five white balls and one black ball. There is nevertheless a difficulty. As Braithwaite recognizes, the Briareus model "will only serve for a probability theory in which every probability is either a proper fraction or 0 or 1". He argues that this constraint is unimportant, on the grounds that continuous distributions "can perfectly well be regarded as convenient techniques for arriving at, or approximating to, probabilities which are rational numbers". But this won't do. It is quite true that in the real world, we consider only finite frequencies. But the motivation for considering a semantics for chance is that we want to take account of ideal possibilities as well as empirical frequencies. Just as we considered a die with a uniform distribution of mass being thrown on a perfectly flat surface, so we may consider an ideal pointer being swung around a circle of unit radius. In that case we want to be able to say that the chance that the pointer ends up between the point 0 and the point one unit along the circumference from 0 is $1/2\pi$. This is just as much a consequence of the laws of mechanics and some mild assumptions as was the corresponding statistical hypothesis about the die. Of course no pointer is ideal; of course we cannot measure distances with infinite precision. Nevertheless if one story makes sense, the other should.

There is a straightforward way of saving the model. Remember that the model is an ideal one, whose business it is to provide semantics for an ideal or theoretical statistical hypothesis. The model as described for the binomial case by Braithwaite consists of a finite number n of urns (corresponding to the parameter n of the binomial distribution) each of which contains a finite number of black and white balls in the proportion p, corresponding to the other parameter of the binomial distribution. For the reasons that Braithwaite adduces in connection with Fisher's infinite urn model, we cannot simply suppose (in order to accommodate irrational measures) that the urns contain an infinite number of balls. But we may nevertheless suppose that the urns grow: we may envisage a sequence of models, each of which involves urns with a finite number of balls. Let us refer to such a sequence as a *Briareus sequence*. At any given point in the sequence the ratio of black to white balls in each urn is determinate and rational. The limit that these ratios approach as we go out in the Briareus sequence may perfectly well be irrational.

To illustrate this model, let us consider a sequence of n trials with our ideal swinging pointer. The ideal long run measure of occasions on which the pointer falls within a distance of one unit along the circumference to the right of the point 0 is $1/2\pi$. The model is as follows: We have a Briareus sequence each member of which consists of a set of n urns. The first member of the sequence is a Briareus model consisting of n urns, each of which contains nine black balls and one white ball. (White corresponds to the event in question.) The second member of the sequence consists of n urns, each of which contains 100 balls, of which fifteen are white. The third member consists of n urns, each of which contains 1000 balls, of which 159 are white; and so on, corresponding to $1/2\pi = 0.159154943\ldots$

Although this model involves a limiting process, it escapes the arguments that Braithwaite offers against the limiting frequency interpretation of probability. The set of urns from which we draw is always finite. We can make the deductions required. The order of observations is irrelevant, since the number of observations is represented by the number of urns (n) which is finite and constant throughout. And the limiting value of the ratio of white to black balls in each urn, as we proceed in the Briareus sequence, is determined by a mathematical rule (corresponding, say, to the decimal expansion of the theoretical parameter involved). Finally, note that we may model any distribution, even a continuous distribution, this way, by having in each urn a number of balls of N different kinds, and allowing both N and the number of balls in the urn to increase as we go along the Briareus sequence.

An actual selection no longer makes sense: we cannot make a selection of a ball from the limit of a sequence of urns. But we can do almost this: we can talk of a *limiting* selection by supposing that each urn in the Briareus sequence of urns has one ball selected from it. Since we are concerned not with the selections themselves, but with properties of those selections, and since those properties proceed to limits in an orderly way, corresponding to the orderly way in which we have designed the sequence, we may speak intelligibly *in the model* of the limiting properties of limiting selections in a way quite analogous to that in which Braithwaite talks of the properties of selections in his one-stage model.

We now have a perfectly good "possible worlds" model giving a semantics for statistical hypotheses. Since it is a *possible* world model, we need not worry about the fact that in the real world there are no infinite sequences of urns. Since we now have a semantics for statistical hypotheses of class III, let us re-examine once more the hypotheses of the first two classes.

Let us consider hypotheses of the second class first. It might be argued that the statistical hypothesis that the wingspan of species X is approximately normally distributed with parameters m and s^2 should be interpreted along the lines we have just suggested for idealized statistical hypotheses. There is a way of doing this. We may suppose that the actual large finite set of individuals of species X constitutes a limiting selection from a Briareus sequence. I do not doubt that we can do this. But it is not clear what the point is, unless we have a general theoretical structure in which these parameters m and s^2 play a role. Recall that in our discussion of statistical hypotheses of type III, the parameters beta, lambda, $1/2\pi$, and $1/6$ were derived from quite general and pervasive theories, and related in certain ways to other constants in those theories. The same *might* be true of the parameters m and s^2. But if that is the case, then we should regard the hypothesis as a hypothesis of type III, and then it would be appropriate to model it in a Briareus sequence. Indeed, this would be necessary if we were to regard the distribution of wingspans as being *theoretically* exactly normally distributed.

But such was not the hypothesis I had in mind when I offered the example. I was imagining a naturalist reporting on the wingspan of species X, not with any ulterior motive, but simply because it is a fact of interest. In that case there seems to be no point in regarding the actual members of species X as a large limiting sample from a Briareus model; on the contrary, if our interest is in the distribution of wingspans among the *actual* individuals of species X, the Briareus model takes us *away* from what we are interested in. If we are to take some action or make some decision to which this distribution of wingspans is relevant, what we want to know is the distribution in the actual world, not the hypothetical distribution in a Briareus world. The Briareus world is irrelevant.

This comes out even more clearly in the case of hypotheses of class I. There is no doubt that we could provide a Briareus sequence that would model the proportion of voters in Lyons that will vote Republican in the next election: we suppose that the total of N voters in Lyons constitutes a limiting selection from a Briareus sequence each member of which has N urns. Our sample then constitutes a selection of m from among those N individuals. But we have no general theory of voting behaviour which gives us the Briareus sequence as a model; and even if we did, what we want to know about is the proportion of the actual N voters from Lyons who will vote Republican.

Thus despite the fact that we have a viable model for theoretical statistical hypotheses – the Briareus sequence model – there are no reasons for, and some reasons against, using this model for hypotheses of classes I and II. For hypotheses of these classes the straightforward, extensional, finite frequency semantics seems more appropriate.

2   To provide a semantics for statistical hypotheses of the three models does not automatically provide us with a way of telling when and whether instances of such hypotheses should be regarded as acceptable, or rejected, or preferred to other instances. We must therefore, as Braithwaite saw clearly, go beyond the semantics of these hypotheses, and provide some conditions which we can apply for accepting and rejecting them.

Braithwaite's approach is two-fold. He first provides for the "empirical meaningfulness" of statistical hypotheses by means of his k-rule of rejection. This is essentially a philosophical use of the idea behind significance testing: we reject, at the k-level, a statistical hypothesis H, when we observe a sample which would lead to the rejection of H less than a proportion k of the time on the assumption that H is true. This rejection is tentative – it may be that we have observed something anomalous. So we apply the same sort of rule again, this time to the statistical hypothesis that no more than a proportion k_1 of such samples will lead to the rejection of H; and so on: 'the possibility of an unending series of empirical tests, none of which reject the statement serves to give, I maintain, an empirical meaning to the statement' (Braithwaite 1953: 162).

Having given this account of the meaning of statistical statements, Braithwaite goes on in the following chapter to offer an account of the grounds on which we may make a choice between two alternative statistical statements. The choice between statistical hypotheses involves the utilities of true and false acceptance and rejection of the hypotheses, which seems undesirable. Of course utilities are involved in the application of statistical knowledge, but one would like to have the weight of evidence bearing on a statistical hypothesis to be analysed in a way that does not depend on the use to which that hypothesis is to be put.

Let us therefore see if the k-rule of rejection, or something like it, can be given a twist which will allow it to determine when and to what degree a hypothesis is supported by given evidence.

Suppose we are concerned with the measure of Bs among As. Braithwaite's k-rule of rejection says to reject the hypothesis that this measure is p, if the observed ratio of Bs to As in a sample of n falls outside the limits of $(p - \sqrt{pq/nk}, p + \sqrt{pq/nk})$. The justification of this rule lies in the Tchebycheff inequality, which says that in employing it we will make false rejections less than a proportion k of the time. Suppose that we have in fact observed n As, of which a fraction f have been observed to be Bs. What is required that we may regard this as grounds for rejecting a particular statistical hypothesis such as: 'The measure of Bs among As is p^*'?

(1) That it not be the case that we have actually observed more than n As. If we have observed a thousand As, we don't want to base our rule of rejection on just a part of that sample.

(2) That we have no reason to suppose that the sample of observed As is atypical. If we are doing a voter survey, we would violate this requirement if we took our sample from only one part of town, for we know perfectly well that any single part of town is

unlikely to be representative of the voting behaviour of the town as a whole.

Both of these requirements are implicit in Braithwaite's treatment of the choice between two statistical hypotheses; both may be regarded as implied by an epistemic principle of total evidence.

Given that we can satisfy this principle of total evidence, we may regard the number $1 - k$ as an index of the cogency of our rejection of the hypothesis that the measure of Bs among As is p^*.

Let us now consider the whole class of hypotheses of the form, 'the measure of Bs among As is p'. Any such hypothesis will be rejected if f lies outside the interval $(p - \sqrt{pq/nk}, p + \sqrt{pq/nk})$. Note that this interval is included in the interval $(p - 1/2\,\sqrt{nk}, p + 1/2\,\sqrt{nk})$. Thus if we reject hypotheses of the form: 'the measure of Bs among As is p', when f lies outside $(p - 1/2\,\sqrt{nk}, p + 1/2\,\sqrt{nk})$, we will even more rarely make erroneous rejections. But this is exactly to reject all these hypotheses with a parameter p falling outside the intervals $(f - 1/2\,\sqrt{nk}, f + 1/2\,\sqrt{nk})$.

Braithwaite warns us at one point (1953: 200) that it may be that *no* statistical hypothesis may be true concerning the measure of Bs among As. For hypotheses of classes I and II, this is clearly not the case: in any non-empty finite class A, there must be some determinate proportion which are Bs. Can it be true for hypotheses of class III? As I argued earlier, hypotheses of class III are hypotheses that are derived from theoretical considerations: they concern such events as throws of a perfect die on an ideal surface, dropping a geometrical line on a geometrical circle, the decay of an atom according to quantum theory, *etc.* If the underlying theory is true, then *some* hypothesis of class III is true. The theory – as in the case of quantum mechanics – may not specify which hypothesis is true. But in either event, if we have grounds for accepting the theory, we have grounds for supposing that some hypothesis of class III is true. So far as evaluating hypotheses is concerned, hypotheses of class III are in the same boat as hypotheses of classes I and II. Despite Braithwaite's warning, we are free to assume that some statistical hypothesis is true.

Return to our example, in which we found we could reject all those hypotheses with a parameter p falling outside the interval $(f - 1/2\,\sqrt{nk}, f + 1/2\,\sqrt{nk})$. This clearly entails that we fail to reject those hypotheses with a parameter falling inside the interval. But since, as we have just argued, some such hypothesis is true, this amounts to *accepting* the disjunction of hypotheses with parameters falling inside the interval. That is to say, it amounts to *accepting* with – to put it in neutral sounding terms – index $1-k$, the hypothesis that the measure of Bs among As lies between $f - 1/2 \sqrt{nk}$ and $f + 1/2 \sqrt{nk}$. (It is perhaps odd to talk of "disjunction", since we are talking about an uncountable disjunction; we can obviously eliminate such talk.)

Since rejection of a set of simple hypotheses at the k-level amounts to acceptance of the complementary set of hypotheses at the $1-k$ level, it is clear that the lottery paradox looms. If we reject a set of S_1 of hypotheses at the k-level, and also reject a set of S_2 of hypotheses at the k-level, on the ground that we will make mistaken rejections no more than a proportion k of the time under similar circumstance, we cannot in general justify the *simultaneous* rejection of the union of S_1 and S_2 on the same grounds.

There are two natural alternatives open to us. The first is to require that in considering the rejection or acceptance of statistical hypotheses, we have a local context in mind. If only the measure of Bs among As is relevant to that context, then the procedure we described earlier is appropriate. If however we are concerned both with the measure of Bs among As and the measure of Cs among Ds, then we must use a procedure that involves the simultaneous rejection of joint hypotheses of the form: The measure of Bs among As is p, and the measure of Cs among Ds is q. The focus on a local context seems clearly to be most compatible with Braithwaite's treatment of the choices between statistical hypotheses.

On the other hand this focus seems clearly at odds with Braithwaite's insistence on generality for scientific laws. He grounds his argument in favour of the Briareus model precisely on the claim that scientific hypotheses should not be concerned with any finite and limited set of objects. Although this extreme requirement seems too strong, the treatment of the decision problem, involving clearly local utilities, seems insufficiently general.

An alternative treatment of the lottery problem is to regard the acceptance and rejection of hypotheses as perfectly general. Thus we accept at the $1-k$ level the hypothesis that the measure of As among Bs is in the interval (p,p'); we accept at the $1-k$ level the hypothesis that the measure of Cs among Ds is in the interval (q,q'); but we do *not* accept at the $1-k$ level the conjunction of these hypotheses – rather we accept at the $1-k$ level the hypothesis that the measure of A–D pairs among B–C pairs, is in the interval (r,r'). In short, we eschew conjunctive, and therefore deductive, closure among the set of statements accepted at the $1-k$ level.

Let us suppose that we can accept, at the $1-k$ level, the hypothesis that the measure of Bs among As is in the interval (p,p'). Under certain circumstances, it will then be appropriate for us to say that the probability that a particular A, say a, is a B is in that same interval. What are the circumstances? Clearly not that "all we know about a is that it is an A" – this is a circumstance that never obtains. The circumstances under which the parameter (p,p') becomes a probability for a particular statement '$B(a)$' are not easy to spell out.[1] To spell them out is precisely to articulate the "principle of total evidence" vaguely referred to earlier.

Suppose that these circumstances can be spelled out. Note that the hypothesis that, whatever be the measure of Bs among As, at least a measure $1-k$ of n-membered subsets of A will exhibit a fraction f of Bs lying in the interval $(p - 1/2 \sqrt{nk}, p + 1/2 \sqrt{nk})$, will be acceptable at any level: it is a mere set-theoretical truth, and the one referred to in justifying the rule of rejection discussed earlier. Put another way, at least a measure $1-k$ of n membered subjects of A will exhibit a fraction f of Bs differing by no more than $1/2 \sqrt{nk}$ from p. Suppose we have examined a sample of n As. If the circumstances which allow us to pass from a measure to a probability obtain for this particular case, then we may conclude that the *probability* is at least $1-k$ that this particular sample exhibits a fraction f of Bs differing by at most $1/2 \sqrt{nk}$ from p, the true measure of Bs among As. This analysis parallels precisely the analysis given previously in terms of the k-rule of rejection. (Recall that in order to *apply* the k-rule of rejection, we also needed a principle of total evidence.)

There are a number of complications that could be considered: $1/2 \sqrt{nk}$ provides an upper bound for the error, and we can do better even in the binomial case by providing a more accurate analysis. The binomial case is one of fundamental importance, but it is not the only one that should be considered. Statistical inference in more complicated cases raises a number of very difficult issues. (See, for example, Seidenfeld 1979.) It is not my concern here to attempt to go into these complications, beyond pointing out that they lurk

---

[1] I have made efforts along these lines in a number of places, particularly my (1961) and my (1974).

along the path of Braithwaite's game-theoretic decision procedure as well as here.

Instead, I shall turn to the structure of these bodies of accepted statements, and to their use in decision-theoretic contexts.

3 We already know that we cannot impose deductive closure on the set of statements accepted at level $1-k$. Since we cannot do that in any case, there is no new problem raised by construing this set of statements as the set of statements whose probability is at least $1-k$, where probability is construed as I have construed it elsewhere. Construed thus, a limited amount of deductive closure is possible: if T is a logical sequence of S, then the probability of T must be at least as great as that of S, and if S is among the accepted set of statements of level $1-k$, so will T be among those statements.

If we construe the set of statements accepted at $1-k$ level as comprised of those whose probability is at least $1-k$, we should ask about the source of that probability. In the examples discussed so far, the frequency part of the probability is simply a set-theoretical truth, and the set of statements to which our "principle of total evidence" is applied will just be our "observations". In general, however, we will be willing to regard justifiable but uncertain statements as part of the evidence relevant to the acceptance of a statement on a lower level. If $1 - k$ *is less than* $1 - k^*$, the set of statements accepted at level $1 - k^*$ may be taken as evidence for statements at level $1-k$. It is often the case, for example, that statistical knowledge acceptable at the $1-k^*$ level can yield an empirical prior distribution which can be combined with data to lead to the acceptance of a statistical hypothesis at the $1-k$ level.

We now have *two* arbitrary constants, k and k^*, to explain rather than one. Although ordinary language provides no obvious precedents, ordinary practice does. In empirical argument it seems quite natural to distinguish between statements whose warrant is such as to allow them to be used as *data*, and statements that are acceptable only in a somewhat weaker sense. Statements may count as data without themselves being "certain". The most common and obvious instances are quantitative data statements in science: We accept such statements as: The length of this rod is l plus or minus d metres, and use them as *data* in the evaluation of physical theories, despite the fact that on the usual theory of errors of measurement, such a statement can be no more than "probable". Similarly, well established statistical statements may serve as data, relative to which we evaluate other statistical statements. We thus suppose that an epistemic situation can be characterized by the pair (k,k^*), where k^* characterizes the level of acceptance corresponding to acceptance *as data*, and k the level of *mere* acceptance. Clearly the warrant for statements in k^* can be questioned; but this involves a shift of epistemic context: we are now construing a different level, $1-k^{**}$, say, as a level appropriate to acceptance as data, and k^* as mere acceptance. In the extreme case, the one most usually addressed by philosophers of induction, the level of acceptance as data is taken to be 1; this is the case discussed by Braithwaite and most other philosophers.

The context of decision is different from the context of inquiry. To decide rationally on the performance of an action A_1, which is appropriate to the state of nature represented by the hypothesis H_1, is not to "accept" H_1. In the ordinary decision problem, mere acceptance has no role to play. But acceptance as evidence does play a role, since we are to base our decision on evidence relevant to the choice between actions. In a typical situation we have two actions, A_1 and A_2; we have two states of nature, represented by two statistical hypotheses H_1 and H_2. We are to acquire evidence E (say, a sample from a population) and base our choice of action on the character of E. Suppose that E may satisfy any one of N exclusive and exhaustive descriptions, $E_1, \ldots, E_N$. Each hypothesis yields a measure on these alternatives. A strategy is a function from the E_i to (A_1, A_2) – that is, it is a plan to adopt action A_1 if any of a certain set of descriptions apply to the evidence, and to adopt A_2 otherwise. We also suppose that we have a utility function yielding values for the four situations: Perform A_1 when H_1 is true, perform A_1 when H_2 is true, perform A_2 when H_1 is true, and perform A_2 when H_2 is true. We represent the utility function by $U(A_i, H_i)$.

Given a hypothesis, H_1, say, we can evaluate any proposed strategy S. Let $U(S_i, H_1)$ be the expected utility of strategy S_i given that H_1 is true, and $U(S_i, H_2)$ be the expected utility of strategy S_i given that H_2 is true. The *maximum* procedure favoured by Braithwaite directs us to adopt that strategy S_j for which the minimum of $U(S_j, H_1)$, $U(S_j, H_2)$ is a maximum: we are to maximize the minimum expectation of our strategy.

Suppose that we can accept as data (at level $1-k^*$) that the (prior) measure of occasions on which H_1 is true is p, and consequently that the measure of occasions on which H_2 is true is $1-p$. This assumes that the decision problem is a member of a class of decision problems about which we have precise statistical knowledge. If we could accept this as data, we could compute the overall expected utility of a certain strategy S_j as $U(S_j) = p(U(S_j,H_1)) + (1-p)(U(S_j,H_2))$. It is clear that in this case the Bayesian approach of maximizing $U(S_j)$ would be preferable to the maximin strategy.

We have seen that it is possible to accept statistical hypotheses. We can even accept them as evidence, since the level of mere acceptance in one context may correspond to the level of acceptance as evidence in another. Clearly in our decision problem, neither H_1 nor H_2 can be regarded as acceptable, but it may well be that a more general statistical hypothesis H^* is acceptable as evidence, and will provide probabilities for H_1 and H_2. These probabilities will in general be intervals, which may be quite broad, $(0,1)$ in the limiting case, or quite narrow – but hardly ever degenerate intervals of the form (p,p). How will this affect our analysis of the decision problem?

We assume that we can compute $U(S_j,H_1)$ and $U(S_j,H_2)$ as before. $U(S_j)$ however must now be computed as an interval:

$$U(S_j) = [\min_{x\varepsilon(p,q)}(U(S_j,H_1).x + U(S_j,H_2).(1-x)),$$
$$\max_{x\varepsilon(p,q)}(U(S_j,H_1).x + U(S_j,H_2).(1-x))]$$

Corresponding to each strategy we now have a pair of numbers. To this set of pairs of numbers, we now apply the maximin procedure: we select that strategy for which the minimum number of the pair is the largest. (In case of ties we select the strategy with the largest maximum expectation.)

It is easy to see that this general procedure selects the same strategy as the maximin procedure in the absence of knowledge concerning the probabilities of H_1 and H_2, and also selects the same strategy as the Bayesian procedure when we have relatively precise probabilities for H_1 and H_2. The latter is obvious, since as the interval (p,q) shrinks, so do the utility intervals $U(S_j)$; in the limit they become degenerate, and the strategies will be ranked according to their expectations under our procedure. The former requires a little argument.

Suppose the probability of H_1 is $(0,1)$. Suppose also that $U(S_j,H_1)$ is greater than $U(S_j,H_2)$. Then

$$min(U(S_j,H_1).x + U(S_j,H_2).(1-x)) = U(S_j,H_2),$$
$$x\varepsilon(0,1)$$

and

$$max(U(S_j,H_1).x + U(S_j,H_2)(1-x)) = U(S_j,H_1)$$
$$x\varepsilon(0,1)$$

Thus, corresponding to S_j is the pair $[U(S_j,H_2), U(S_j,H_1)]$. If $U(S_j,H_2)$ is greater than $U(S_j,H_1)$, we will have the same pair but in reverse order. But these are exactly the same pairs we considered in our initial discussion of the maximin procedure, and that procedure will lead to the same result we were led to there.

4   We have not examined the principle of total evidence which will allow us to pass from statistical knowledge to corresponding probabilities. This is a surprisingly complicated matter.[2] But no one, I think, doubts that there *are* circumstances under which knowledge of the measure of As among Bs, together with knowledge that x is a B, can legitimately yield a corresponding probability for 'x is an A'. This is the only principle we have (implicitly) employed in the foregoing; we have not introduced either "subjective" probabilities, or probabilities corresponding to logical measures on the set of sentences in a language, both of which Braithwaite persuasively inveighs against. It is interesting that no more than a principle of total evidence that will support a principle of direct inference from general statistical knowledge to particular cases is needed to yield insight both into the grounds for the acceptance of statistical statements, and into the principles for the application of statistical knowledge to decision problems.

University of Rochester

REFERENCES

Braithwaite, R. B. 1953. *Scientific Explanation*. Cambridge.
Kyburg, H. E. 1961. *Probability and the Logic of Rational Belief*. New York.
Kyburg, H. E. 1974. *The Logical Foundations of Scientific Inference*. Reidel.
Kyburg, H. E. 1975. The use of probability and the choice of a reference class. *Minnesota Studies in the Philosophy of Science Vol. 6*, ed. G. Maxwell and R. M. Anderson, pp. 262–94. Minneapolis.

[2] See, for example, my (1975); Levi (1977) and my response (1977); Levi (1978) and Seidenfeld (1978).

Kyburg, H. E. 1977. Randomness and the right reference class. *Journal of Philosophy* **74**, 501–21.

Levi, I. 1977. Direct inference. *Journal of Philosophy* **74**, 5–29.

Levi, I. 1978. Confirmational conditionalization. *Journal of Philosophy* **75**, 730–7.

Seidenfeld, T. 1978. Direct inference and inverse inference. *Journal of Philosophy* **75**, 710–30.

Seidenfeld, T. 1979. *Philosophical Problems of Statistical Inference*. Dordrecht.

10 *How is it reasonable to base preferences on estimates of chance?*

R. C. JEFFREY

As Richard Braithwaite (1965) did in Jerusalem, 15 years ago, here I take it for granted that it is reasonable to base action on estimates of chance, but where he asked 'Why?', I now return to the question 'How?', since the answer I floated in Jerusalem (Jeffrey 1965a) now seems questionable.

The answer in question is, 'Rank possible courses of action according to their conditional subjective expectations of utility', *i.e.* according to their "desirabilities", in my Pickwickian sense of that term. And the question about this answer is, 'Does it take proper account of the difference between actions as symptoms and as causes of states of affairs that we act to promote or avert?' To this question the answer seems to be 'No' because of certain counter-examples (Nozick 1969, Gibbard and Harper 1978, Lewis 1979) that would be more troublesome if less *outré*. But there are homely specimens of the type, *e.g.* the one presented here in **1**. Then the aetiology *vs.* prognosis problem about action is no mere fantasy. Still, the situation would be clearer if counterexamples could be found in which that problem attaches to events other than actions, or if one could see why counterexamples of that sort cannot be forthcoming. A failed attempt to find such a counterexample is set forth in **2**. Finally, **3** is an incomplete, inconclusive survey of answers to the main question above.

1 *Running.* Shall I take up running? I would, for my health, if I did not loathe exercise. But should I? Think! Let H, E, and L be the propositions (H) that I will be in good health 3 years hence, (E) that I exercise regularly for the next 3 years, and (L) that I loaf instead, for the next 3 years. For me, loafing dominates exercising (relative to H

vs. −H) in the sense that I prefer *LH* to *EH*, and I also prefer *L −H* to *E−H*. The degrees of those preferences are shown in the table of desirabilities: $d(EH)=8$, $d(E−H)=0$, *etc*. And on the basis of statistics and other information about myself and people whom I take to be like me in the relevant respects, my credences in *H* conditionally upon *E* and upon *L* are $c(H/E)=0.9$ and $c(H/L)=0.6$, as shown in the table of conditional credences. (Perhaps among people whom I take to be relevantly like me, 90 per cent of the 3-year exercisers are healthy, but only 60 per cent of the 3-year loafers are.)

	H	−H
L	9	1
E	8	0

Desirabilities

	H	−H
L	0.6	0.4
E	0.9	0.1

Conditional credences

Now desirabilities obey the rule

(1) $d(A) = d(AH)c(H/A) + d(A−H)c(−H/A)$ if $c(A) \neq 0$

where *A* and *H* are any propositions, *e.g.* where $A=E$ and when $A=L$, with *H*, *E*, and *L* as above. Then it is a matter of calculation to verify that $d(E) = 7.2$ while $d(L) = 5.8$, so that if preference goes by desirability (by conditional expected utility) then my credences and my desirabilities concerning *EH*, *LH*, *E−H*, and *L −H* imply that, all things considered, I prefer exercise to loafing.

Perhaps I do, and perhaps I should, *e.g.* I should if the difference between my conditional credences $c(H/E)=0.9$ and $c(H/L)=0.6$ is due to my belief in the healthy effects of exercise as against loafing. But suppose I think there are no such effects. Suppose I think that among people like me, it is those with healthy constitutions who find exercise attractive, while the loafers loaf because of their unhealthy constitutions. Suppose I think that exercising is a symptom of a constitution, the possessor of which would be healthy even if he loafed, while loafing is a symptom of a different constitution, the possessor of which would be unhealthy even if he exercised. I could think that and still have the conditional credences that are shown above; and if I thought that, I would be a fool to exercise, for I think it would do no good. I would then be well advised to loaf, for I prefer *L* to *E* whatever my fate (*H* or −*H*), and my choice between *L* and *E* has no influence on my fate.

In this example, neither of the extreme positions seems tenable:

neither the position that my choice between L and E is purely efficacious in relation to the issue between H and $-H$, nor the position that it is purely symptomatic (of a constitution that is purely efficacious and is uninfluenced by my choice). In other examples, the first of these pure positions is plausible, *e.g.* it is plausible that my choice between smoking and abstaining is purely efficacious in relation to H *vs.* $-H$; but such examples are not counterexamples to my Jerusalem proposal. The other sort of pure position is the one found in the bizarre counterexamples, *e.g.* Newcomb's Problem (Nozick 1969). There seem to be few believable examples in which this second pure position is tenable, *i.e.* in which choice is purely symptomatic. (Lewis (1979) argues that The Prisoner's Dilemma is one.) But perhaps pure cases are not needed. Perhaps the present example is persuasive as a counterexample in its plausible, mixed form, where I take my choice between L and E to be partly efficacious, but partly (and in large part) symptomatic.

2 *Optional stopping.* The running example is persuasive to the extent that I can be thought to look to my own choice between exercising and loafing for prognosis that is in large part unmediated by hope of therapy or of prophylaxis. But to the extent that my reasoning can be seen in that way, it is difficult to see it as an example of decision making instead of as an attempt to gain information that might be useful in making some other, genuine decision, *e.g.* a decision about undertaking a long-term project that demands good health for success. Are there no counterexamples in which the thesis that preference always goes by desirability confounds prognosis with aetiology in relation to events that are not acts? Such counterexamples would be especially compelling, but I have not been able to find one. Or if I have found one, I have failed to recognize it. Here is a case in point: an example that turns on the indistinguishability *from the outside* of two urn processes: *Pólya* and *Bayes*.

The Pólya process involves sampling with *double* replacement from an urn that initially contains one red ball and one black one. A *trial* of this process is a matter of randomly drawing a ball from the urn and then replacing it, together with one more ball of its colour. This means that after t trials there will be $2+t$ balls in the urn. A *success* is a trial on which a red ball is drawn. After s successes, there will be $1+s$ red balls in the urn. Then after t trials, of which s have been successes, the ratio of red balls to all in the urn will be $1+s:2+t$, and that should be one's initial degree of belief in success on a trial after the tth, conditionally upon a particular s successes and $t-s$ failures in the first t trials: with S_k meaning that *the kth trial is a success*, we have

(2) $c(S_{t+i}/\pm S_1 \ldots \pm S_t) = \dfrac{1+s}{2+t}$ *if there are s plusses and $i > 0$.*

Now consider the following two arrangements (*sc.* propositions A, B, that those arrangements are in effect).

(A) *That one gets £1 if S_1 is true, and pay £1 if S_1 is false.*

(B) *Ditto for S_{u+1}, where u is the first t for which $2s > t$.*

The arrangements (to make A true, or B) are to be made before the first trial. We assume that the desirabilities of A and of B are simply subjective gains, and with no discounting of future gains as against immediate ones. Now as the ratio of red balls to all in the urn is 1:2 just before trial 1, and is greater than that just before trial $u+1$, B is preferable to A.

It is a surprising and instructive fact (de Finetti 1975: §11.4.4) that from the outside the Pólya process is indistinguishable from the *Bayes (–Laplace–Carnap)* process, *viz.*, sampling with single replacement from an urn of fixed, unknown composition, where initially the ratio r of red balls to all in the urn is thought equally likely to lie in either of any two equal subintervals of the unit interval (*i.e.* where the prior distribution of r is *uniform*). Now if there are s plusses we have $c(\pm S_1 \ldots \pm S_t) = \int_0^1 r^s(1-r)^{t-s}\, dr$, and therefore

(3) $c(S_{t+i}/\pm S_1 \ldots \pm S_t) = \dfrac{\int_0^1 r^{s+1}(1-r)^{t-s}\, dr}{\int_0^1 r^s(1-r)^{t-s}\, dr} = \dfrac{1+s}{2+t}$ (*s plusses*)

just as in the Pólya case.

Note that when all trials are successful (when $s=t$), both (2) and (3) yield Laplace's rule of succession,

(4) $s(S_{t+i}/S_1 \ldots S_t) = \dfrac{1+t}{2+t}$

Note, too, that the credence function determined by (2) and by (3) is Carnap's (1945) function c^* for a language with just one monadic primitive predicate.

The two processes are *indistinguishable from the outside* in the sense that, initially, one's credences ought to be the same, regarding (finite or infinite) Boolean combinations of propositions S_i concerning outcomes of trials, no matter which process one thinks one is about to watch; and, therefore, observation of outcomes can give no clue about what is going on behind the scenes ("on the inside"). But if we extend the domain of definition of c to include statements of form $P(a,b,t)$ which say that *the ratio r of red balls to all in the urns is in the interval from a to b just after the tth trial*, then differences will appear. Thus in the Bayes case but not in the Pólya case we should have $c(P(a,b,t)/P(a,b,0)) = 1$ for $t=1,2, \ldots$ and $0 < a < b < 1$. But in either case, if credence in the conjunction $H.P(a,b,t)$ is positive we should have

(5) $a < c(S_t/H.P(a,b,t)) < b$, *where*
      H is a Boolean compound of $S_1, \ldots, S_t$ and
      $P(a,b,t)$ = *the proposition that $a \leqslant r < b$ just after t.*

In these two processes the propositions $P(a,b,t)$ specify intervals within which the objective *chance* of success on the next trial lies, just after the tth, and (5) indicates how it is reasonable to base credence on estimates of chance in such cases.

If preference always goes by desirability, estimates of chance influence preference only by way of their influence on credences, so that one's preference between A and B should be the same if one thought one was about to witness a Bayes process instead of a Pólya process: B should be preferred to A in either case. But it is easy to imagine that, in the Bayes case, one should be indifferent between A and B, so that here we have the desired, conclusive sort of counterexample.

But one tempting route to that conclusion is simply a mistake. It goes like this. 'Sooner or later, statistical fluctuations will produce a sequence of u trials in which more than half are successes. Then initially, credence should be 1 in the proposition that the condition in B will be met sooner or later. Then, when it is met, the fact that it is met tells us nothing about the fixed, unknown composition of the urn in the Bayes case. Thus in the Bayes case, the $(u+1)$th trial is on a par with the first, and there is nothing to choose between A and B except for immediacy, which we have agreed to ignore.' That is wrong because initially it is far from sure that the condition in B will be met: initial credence in the proposition that, eventually, more than half of the trial will have been successes, should be strictly between 0 and 1. "Statistical fluctuations" will almost surely produce more than s successes for each finite s, but the rate of production of successes may well be slow enough so that s/t never exceeds $\frac{1}{2}$, as t and s increase without bound. (The issue between Armitage and Savage in Savage *et al.* 1962: 72–3 is of this sort.)

Then, in the Bayes case, the coming true of the condition in B is a genuine sign that success on the next trial is more credible than success on the first trial was, initially, and therefore B is genuinely to be preferred to A in this case, too (although for a different reason than in the Pólya case, if reasons are given in terms of what is thought to be going on behind the scenes). So I see it, anyway, and so I do not see this as a counterexample to the thesis that desirability always corresponds to preferability. But the running example is enough, if it withstands scrutiny.

3 *Four answers.* Let us now review some of the possible answers to the question that forms the title of this piece. In discussing them I shall distinguish among *credence* (subjective probability), *propensity* (chance, objective probability), and *relative frequency* (which is no kind of probability at all, although it may strongly affect one's credence and may be valuable evidence concerning propensity). But I shall not argue here for the tenability of these three dualisms. Having used 'c' for credence, I cannot use it for chance as well: hence, 'p' for propensity, *i.e.* chance.

Following a suggestion of Robert Stalnaker's, Gibbard and Harper (1978) propose to replace my identification of preference with desirability, d, as in (1), by a variant in which conditional credences $c(H/A)$ and $c(-H/A)$ are replaced by unconditional credences $c(A \rightarrow H)$ and $c(A \rightarrow -H)$ in counterfactual conditionals:

(6) $e(A) = d(AH)c(A \rightarrow H) + d(A-H)c(A \rightarrow -H)$

(Gibbard and Harper follow Lewis (1973) in using a box followed by an arrow where I use a simple arrow for the counterfactual conditional connective.) The suggestion works well in the pure cases, *e.g.* where I am sure that A is purely a symptom, not a cause of H or of $-H$, then $c(A \rightarrow H) = c(H)$ and $c(A \rightarrow -H) = c(-H)$, and, if this is the case for all available acts A, then the dominant act will have the highest e value. (Here it is essential that the counterfactuals be given *nonbacktracking* readings, *e.g.* '$E \rightarrow H$' is to be read as a claim about what would happen if I, with my actual constitution, were to exercise, and not as a claim about how things would have fallen out if I had been born the sort of person who exercises.)

My present inclination is to approach the problem from an eclectic point of view that (I think) many statisticians find natural. Recall that Savage (1954) viewed acts as mappings a of what one might call 'possible worlds' $w \varepsilon W$ into an abstract set of items called 'consequences'. It is propositions (sets of possible worlds) that have credences, and it is consequences that have utilities. Acts *a* have *expected utilities* $\tilde{u}(a)$ that correspond to their positions in the agent's preference ranking:

(6) $\tilde{u}(a) = \mathbf{E}(u(a(\cdot)),c,W) = \Sigma_{w \varepsilon W} u(a(w))c(\{w\})$ (*Savage*)

where the sum at the right is what the expectation $\mathbf{E}(u(a(\cdot)),c,W)$ comes to in the discrete case, where W is countable. (In general, the expectation is an integral.) In my 1965 scheme, Savage's acts *a* are replaced by propositions A, and consequences play no explicit role. Act-propositions have conditional utilities:

(7) $d(A) = \mathbf{E}(u,c,A) = \Sigma_{w \varepsilon W} u(w)c(\{w\}/A)$ (*Jeffrey* 1965a)

In these terms, the Gibbard–Harper proposal might be this:

(8) $e(A) = \mathbf{E}(u,c(A \to \cdot),W) = \Sigma_{w \varepsilon W} u(w)c(A \to \{w\})$

                                               (*Gibbard–Harper?*)

(The query in (8) indicates that this is speculation on my part. Gibbard and Harper (1978) say nothing about the relationship between the set function e (their U) and the point function u. Observe that the integral makes sense only if the set function $c(A \to \cdot)$ is additive, *i.e.* only if $c(A \to (B \vee C)) = c(A \to B) + c(A \to C)$ whenever B and C are incompatible, conditionally upon A.)

One striking feature of Savage's scheme is that, in it, states are independent of acts. I think he meant that independence to be both probabilistic and causal. That feature was common ground among statistical decision theorists of a general Bayesian stripe in 1965, when I proposed dropping it in order to make it possible to directly consider decision problems in which acts are undertaken with a view to making the more desirable states of nature more probable than they would otherwise be, as when I take up running because I think it conducive to health. To treat that sort of problem, Savage had to resort to what strikes me as a clumsy artifice: he had to cook up states of nature that were independent of acts (Savage 1954, §2.5; Jeffrey 1977b). Gibbard and Harper take off from that aspect of Savage's scheme, analysing his artificial act-independent states by means of counterfactuals.

A different, equally striking feature of Savage's scheme is that, in it, acts are fully deterministic: in each possible world w, each act a has a perfectly definite consequence $a(w)$. That feature was peculiar to Savage's scheme, reflecting his thoroughgoing subjectivism. It was by no means common ground among Bayesian statistical decision theorists in 1954 or in 1965, nor is it today. Statistical decision theorists are far more likely to represent the relationship between acts and consequences as mediated by states of nature that are objective probability distributions over consequences. (Commonly, statisticians will identify objective probability with relative frequency, or with what relative frequency would be if experiments were repeated *ad infinitum*, but that strikes me as nonsense for reasons given in 'Mises redux' (Jeffrey 1977a), where I propose the sort of mixed Bayesianism that I have in mind here.) Here is a schematic account of how that would go.

Let 'p' range over a set P of probability measures that contain all of those that the agent thinks might be the true propensity, and let c be his actual credence function, which I take to be defined both on the algebra of subsets of W that all members of p are defined on, and also on a suitable algebra of subsets of P. Now, relative to any particular p, the conditional expected utility of act A is $\mathbf{E}(u,p,A)$, and so, given the agent's opinion about where the true p is likely to lie in P, his overall preference concerning A will be given by his subjective expectation $\mathbf{E}(\ ,c,P)$ as p ranges over P:

(9) $U(A) = \mathbf{E}(\mathbf{E}(u,p,A),c,P)$ (*Mixed Bayesianism*)

This account is schematic rather than general, *e.g.* because "the" algebra of subsets of P associated with the outer expectation must be chosen to fit the case at hand, in a way that may require considerable tact, ingenuity, and luck, if (9) is to make sense and be useful in that case.

Elsewhere (Jeffrey 1980) I have attempted a survey of the present state of inquiry into the main question, but the situation is fluid, and as I write much of the relevant literature is in press (Skyrms 1980) or still in draft form (Howard Sobel) or still unwritten (David Lewis).

Princeton University

REFERENCES

Braithwaite, R. B. 1965. Why is it reasonable to base a betting rate upon an estimate of chance? In *Logic, Methodology, and Philosophy of Science*, ed. Y. Bar-Hillel, pp. 263–73. Amsterdam.

Carnap, R. 1945. On inductive logic, *Philosophy of Science* **12**, 72–97.

de Finetti, B. 1975. *Theory of Probability*. London.

Gibbard, A. & Harper, W. L. 1978. Counterfactuals and two kinds of expected utility. In *Foundations and Applications of Decision Theory*, ed. C. A. Hooker, J. J. Leach & E. F. McClennen, pp. 125–62, Dordrecht.

Jeffrey, R. C. 1965a. New foundations for Bayesian decision theory. In *Logic, Methodology, and Philosophy of Science*, ed. Y. Bar-Hillel, pp. 289–300. Amsterdam.

Jeffrey, R. C. 1965b. *The Logic of Decision*. New York.

Jeffrey, R. C. 1977a. Mises redux. In *Basic Problems in Methodology and Linguistics*, ed. R. E. Butts & J. Hintikka, pp. 213–22.

Jeffrey, R. C. 1977b. Savage's omelet. In *PSA 1976*, volume 2, ed. F. Suppe & P. D. Asquith, pp. 361–71. East Lansing, Michigan.

Jeffrey, R. C. 1980. Choice, chance, and credence. In *Philosophy of Logic*, ed. G. H. von Wright & G. Fløistad. Nijhoff.

Lewis, D. K. 1973. *Counterfactuals*. Cambridge, Mass.

Lewis, D. K. 1979. Prisoners' dilemma is a Newcomb problem, *Philosophy and Public Affairs* **8**, 235–40.

Nozick, R. 1969. Newcomb's problem and two principles of choice. In *Essays in Honor of Carl G. Hempel*, ed. N. Rescher, pp. 114–46. Dordrecht.

Savage, L. J. 1954. *The Foundations of Statistics*. New York.

Savage, L. J. *et al.* 1962. *The Foundations of Statistical Inference*. London.

Skyrms, B. 1980. *Causal Necessity*. New Haven, Conn.

11 Uses of Bayesian probability models in game theory[1]

JOHN C. HARSANYI

1 *Objective and subjective probabilities and rational behaviour.* Operationally, we can distinguish between *objective* and *subjective* probabilities as follows. Some probabilities are assigned the same numerical values by virtually all expert observers: for example, virtually all experts will agree that a fair coin will show either side with the same probability 1/2; or that a fair die will show any one of its six sides with the same probability 1/6; *etc.* Other probabilities will typically be assessed differently by different people: for example, different experts on horse racing will usually assign different numerical probabilities to a particular horse coming in first in a given race. Probabilities of the former type may be called objective probabilities, while those of the latter type may be called subjective probabilities.

Objective probabilities can be regarded as objective *structural properties* of certain physical (or biological or psychological or social) systems; and they manifest themselves in the actual statistical behaviour of these systems, *i.e.* in the long-run frequencies of various alternative outcomes (*e.g.* in the fact that the long-run frequency of HEADS in tossing a fair coin tends to be very close to 50 per cent). In contrast, subjective probabilities express *psychological attitudes* (expectations) held by various observers and decision makers, rather than structural properties of the objects observed and acted on.

[1] The author wants to thank the National Science Foundation for supporting this research through Grant SOC 77–06394 to the Center for Research in Management Science, University of California, Berkeley. He also wants to express his gratitude to the Center for Interdisciplinary Research (Zentrum für interdisziplinäre Forschung) of the University of Bielefeld, West Germany, for providing ideal working conditions for him during the academic year 1978–9.

189

The objective probabilities characterizing the behaviour of many physical systems can be ascertained only by detailed statistical observations. But for other physical systems these probabilities can often be predicted on the basis of theoretical considerations, more particularly on the basis of possible *symmetry* properties displayed by some of these systems (as in the case of a fair coin, or a fair die, *etc.*).

The distinction between objective and subjective probabilities gives rise to the following threefold classification of *individual* decision problems (*i.e.* of problems of *individual* rational behaviour). We say that a given decision maker acts under *certainty* when he can predict the actual outcome (*i.e.* the actual consequences) of any action he can take. We say that he acts under *risk* when he can at least predict the objective probabilities associated with different possible outcomes. Finally, we will say that he acts under *uncertainty* when some or all of these probabilities are unknown to him (or when they do not even exist as well-defined objective probabilities).

Since the subject of this paper will be the uses of subjective probabilities in game theory, it may be helpful to say something about the relationship between such *individual* decision problems and *game-theoretical* decision problems. (I will make additional comments on this relationship in Section **3** below.) Decision problems in general can be divided into those of *individual* decision, and those arising in a *social* setting. The latter in turn can be divided into *moral* decision problems and *game-theoretical* ones. A *moral* decision problem is a problem of how the different individual members of society can best promote the *common good*, *i.e.* can best promote the common interests of society as a whole. In contrast, a *game-theoretical* decision problem is a problem of how two or more rational individuals, with possibly very dissimilar interests, can best promote their *own* interests *against*, or possibly in cooperation *with*, one another. (Note that game situations involving conflicts of interest can arise not only among *egoists*, each of them trying to promote his personal self-interest, but rather also among *altruists*, trying to promote some altruistic objectives – as long as the latter are trying to promote different and more or less incompatible altruistic objectives.)

Of course, all three categories of individual, moral, and game-theoretical, decision problems come under the general heading of problems of *practical* rationality, *i.e.* of problems of rational choice among alternative *actions*; and thus contrast with problems of *theoretical* rationality, *i.e.* with problems of rational choice among alternative theoretical *beliefs*. (The problems of theoretical rationality in turn comprise both problems of *deductive*, and of *inductive*, rationality.) Thus, our classification of various problems of rationality can be summarized like this:

(A)   *Theoretical* rationality.
    (A1)   *Deductive* rationality.
    (A2)   *Inductive* rationality.
(B)   *Practical* rationality (rational behaviour).
    (B1)   *Individual* rational behaviour.
        (B1a)   Under certainty
        (B1b)   Under risk
        (B1c)   Under uncertainty.
    (B2)   *Moral* rationality.
    (B3)   *Game-theoretical* rationality.

Leaving the problems of theoretical rationality on one side,[2] the rationality criteria (decision rules) appropriate for the various specific subcases of practical rationality can be described as follows. Already classical economic theory has established the fact that rational behaviour under *certainty* (case B1a) can be characterized as *utility maximization* (see, *e.g.* Debreu 1959: 55–9). Modern Bayesian decision theory has extended this result by showing that rational behaviour under *risk* and under *uncertainty* (cases B1b and B1c) is equivalent to *expected-utility maximization*. More particularly, in the case of risk, where all relevant probabilities are known to the decision maker, he must maximize an expected-utility expression that uses these known *objective* probabilities as probability weights; whereas in the case of uncertainty he must use his own *subjective* probabilities in lieu of the unknown objective probabilities. (For a discussion of the rationality axioms actually needed to establish the expected-utility maximization theorem, see Harsanyi 1978.)

In the case of *moral* rationality (case B2), I have tried to show that the appropriate rationality criterion is maximizing the *average utility*

---

[2] Modern logic provides a very specific and very convincing characterization of deductive rationality. In contrast, in my opinion, we still lack a satisfactory theory of inductive rationality – though it *might* be possible to develop such a theory by suitable extension of Bayesian decision theory.

level of all individuals in society (Harsany 1953; 1955; 1976, Chapters I to V; 1977a, Chapter 4; for a brief summary of my theory, see Harsanyi 1977b). Finally, in the case of *game-theoretical* rationality, the required rationality criteria are presumably provided by the various game-theoretical solution concepts. (For a critique of the solution concepts of classical game theory, and for suggestion of an alternative solution theory, see Harsanyi 1977e.)

Even though the five classes of practical decision problems (the problems of rational behaviour under certainty, risk, and uncertainty, as well as the problems of morality, and of game-theoretical rationality) all represent quite distinct analytical problems, in actual fact there has been a good deal of interaction among individual decision theory, moral theory, and game theory. For example, von Neumann and Morgenstern (1947) already found it useful to provide an axiomatic analysis of rational behaviour under *risk* in order to clarify certain game-theoretical problems. My own work on moral theory (see my papers quoted above) has used the same theory of rational behaviour under risk as a foundation for a utilitarian theory of morality. In this paper I now propose to discuss how the Bayesian theory of rational behaviour under *uncertainty* has been used to solve certain game-theoretical problems.

2 *Bayesian probability models for games with incomplete information.* Classical game theory uses probabilities for two main purposes. One is to characterize *random moves*. The other is to characterize various classes of randomized strategies, such as *mixed strategies* (both individually randomized and jointly randomized mixed strategies) and *behaviour strategies*. In both cases, the numerical probabilities employed are interpreted as *objective* probabilities, and are assumed to be generated by suitable random devices with fully known statistical properties.

The first major use of *subjective* probabilities in game theory has been, I believe, in analysing games with *incomplete information* (Harsanyi 1967–8). A notable feature of this approach has been that, whereas the games to be studied are originally described in terms of certain *subjective* probability distributions (which express the players' expectations about some game parameters unknown to them), in the actual course of the analysis these subjective probability distributions are replaced by *objective* probability distributions of the same mathematical forms.

To describe the nature of the problem, I have to discuss two related, but actually quite different, taxonomic distinctions: that between games with *perfect* and with *imperfect* information, and that between games with *complete* and with *incomplete* information. A game is said to involve *perfect* or *imperfect* information according as the players do or do not have full information at every stage of the game about all *past moves*, including all personal moves made by the various players, and all random moves decided by chance. On the other hand, a game is said to involve *complete* or *incomplete* information according as all players do or do not have full information about the very *nature* of the game as specified by the extensive form (*i.e.* the game tree), or at least as specified by the normal form (*i.e.* the payoff matrix).

Reflection will show that almost all real-life social situations are games with incomplete information. In most cases the players will have less than full information about each other's *payoff functions*. (First of all, they may be unable to predict the physical outcome to be produced by various possible strategy combinations. Secondly, they may not know other players' preference rankings among different possible physical outcomes. Finally, even if they do know these preference rankings (and, therefore, know the other players' ordinal utility functions), they may not know the other players' attitudes to risk taking (*i.e.* may not know the other players' cardinal utility functions).) Likewise, the players may not know the specific *strategies* available to other players. Finally, they may not know how much *information* the various players have about the other players' payoff functions and strategy sets.

Yet, in spite of the great importance that games with incomplete information have in empirical applications – as well as from a theoretical point of view – classical game theory was totally unable to analyse such games. (But it had no difficulty in dealing with games involving either *perfect* or *imperfect* information, as long as these were games with *complete* information.)

The problem, however, can easily be overcome with the help of Bayesian decision theory.

(i) Let G be an n-person game with incomplete information in normal form. We can always model the players' incomplete information by assuming it to be incomplete information about the *other* players. More specifically, we can use the following model. The role of each player i in the game can be played by any one of a number of individuals with different personal characteristics, different power positions, different resource endowments, *etc.* These different individuals will be called the possible *types* (or *subplayers*) of player i. They will be numbered as type 1, 2, . . ., k_i, . . ., K_i, of player i. Incomplete information can now be modelled by the assumption that, whereas each player i will know his *own* actual type (*i.e.* he *will* know the number k_i), he will not know the *other* players' actual types (he will *not* know the numbers k_1, . . ., k_{i-1}, k_{i+1}, . . ., k_n).

In a game with complete information, the payoff function of any given player i can be written as $U_i(s_1, \ldots, s_i, \ldots, s_n)$, indicating that his payoff will depend on the strategies $s_1, \ldots, s_i, \ldots, s_n$ of players $1, \ldots, i, \ldots, n$. But, in our incomplete-information game, each player i's payoff function will have to be written as $U_i(s_1, \ldots, s_i, \ldots, s_n; k_1, \ldots, k_i, \ldots, k_n)$, to indicate that in general his payoff may depend, not only on the various players' strategies, but also on their actual types.

(ii) Assuming that all players follow the rationality postulates of Bayesian decision theory, each player i will express his expectations about the actual values of the unknown variables $k_1, \ldots, k_{i-1}, k_{i+1}, \ldots, k_n$ in the form of a *subjective* joint probability distribution $P_i = P_i(k_1, \ldots, k_{i-1}, k_{i+1}, \ldots, k_n)$. For more convenient notation, I will write $\overline{k_i} = (k_1, \ldots, k_{i-1}, k_{i+1}, \ldots, k_n)$, so that P_i can be written as $P_i = P_i(\overline{k_i})$.

Of course, the other players in general *will not know* the subjective probability distribution P_i entertained by player i. Within our model this fact can be expressed by the assumption that the different types of player i would in general entertain *different* subjective probability distributions P_i. (As in our model the other players will not know what player i's type actually is, this assumption ensures that they will not know the distribution P_i either.) To indicate this dependence of P_i on player i's actual type, we will denote the subjective probability distribution used by any given type k_i of player i as $P_i(\overline{k_i}|k_i)$. We will also use the notation

(2.1) $\Pi_i = \{P_i(\overline{k_i}|1), P_i(\overline{k_i}|2), \ldots, P_i(\overline{k_i}|K_i)\}$

for $i = 1, \ldots, n$.

Thus, Π_i is the set of *all* subjective probability distributions that would be used by the various types $1, 2, \ldots, K_i$ of a given player i.

Full characterization of the game will now obviously require, not only specification of game G itself, but also specification of the sets

$\Pi_1, \ldots, \Pi_n$. Accordingly, we will write $G° = \{G; \Pi_1, \ldots, \Pi_n\}$. Thus, $G°$ is simply game G, as enriched by information about the subjective probability distributions that would be used by the various possible types of every player.

(iii) The next step of our analysis will involve construction of a new game G^*, similar to game $G°$, except that all *subjective* probability distributions $P_i(\overline{k_i}|k_i)$ entertained by the various types k_i of any given player i will be replaced by *objective* probability distributions $R_i(\overline{k_i}|k_i)$ of the same mathematical forms.

To be more specific, in many social situations it will be natural to choose the subjective probability distributions $P_i(\overline{k_i}|k_i)$ in such a way that all of them will be obtainable as *conditional* probability distributions

$$(2.2) \quad P_i(\overline{k_i}|k_i) = P_i(k_1, \ldots, k_{i-1}, k_{i+1}, \ldots, k_n|k_i)$$
$$= R(k_1, \ldots, k_{i-1}, k_{i+1}, \ldots, k_n|k_i),$$

derived from the *same* basic joint probability distribution

$$(2.3) \quad R = R(k_1, \ldots, k_{i-1}, k_i, k_{i+1}, \ldots, k_n).$$

If this is the case then the desired game G^* can be obtained by using the following simple model.

At the beginning of the game, nature conducts a "lottery" in order to choose the types $k_1, \ldots, k_n$ who will play the roles of players $1, \ldots, n$ in the game. This lottery will be governed by the joint probability distribution R (which, therefore, will have the nature of an *objective* probability distribution). This distribution R will be known to all players (and to all types of every player). Each player i will also know his own type but will be ignorant of the *other* players' actual types. Yet, since he will know the probability distribution R and will know his own type k_i, his information about the player types likely to represent the other players will correspond to the *conditional* probability distribution $R(k_1, \ldots, k_{i-1}, k_{i+1}, k_n|k_i)$. In view of (2.2) this *objective* conditional probability distribution will have the same mathematical form as the corresponding *subjective* probability distribution $P_i(k_1, \ldots, k_{i-1}, k_{i+1}, \ldots, k_n|k_i)$ occurring in game $G°$ did.

Of course, in other social situations we may have to postulate subjective probability distributions P_i which *cannot* be derived from any given basic distribution R as conditional distributions. In such cases, a more complicated construction is required in order to obtain a suitable game G^*. (In particular, instead of having one "lottery" at the beginning of the game, we may need several "lotteries" at the end.) But it can be shown that it is *always* possible to construct a game G^* in which the subjective probability distributions $P_i(\overline{k_i}|k_i)$ are replaced by objective probability distributions of the same mathematical forms (see Harsanyi 1967–8: 486–502).

(iv) I now propose to argue that, once a game G^* with the desired properties has been constructed, this game will be game-theoretically *equivalent* to game G°. This follows from the general principle that Bayesian decision makers (including the players of a game if they follow the Bayesian rationality postulates) will always act on any given set of numerical probabilities in *exactly the same way*, regardless of whether these probabilities are *objective* probabilities based on the best information available to them, or are their own carefully chosen *subjective* probabilities (when the information available does not include a knowledge of the relevant objective probabilities).

Yet, this means that the problem of analysing a game with *incomplete* information, G (or G°), has been reduced to the problem of analysing a game with *complete* (even though *imperfect*) information, G^*. (G^* is a game with complete, yet with imperfect, information because the players' original ignorance about some basic parameters of the game has been replaced by an ignorance about the outcomes of certain random moves or "lotteries".)

These modelling techniques, involving an introduction of suitably chosen random moves, have opened up a completely new dimension of flexibility for game theory in the study of game situations. We now can produce games with any desired distribution of knowledge and of ignorance among the various players, and can conveniently study how alternative informational assumptions will change the nature of any given game. We can also study how any given player can *infer* some pieces of information originally denied to him, by observing the moves of some other players who already possess this information, and whose moves may in fact express this information. We can also investigate how each player can optimally *convey* information to some other players, or can optimally *withhold* information from them, in accordance with his own strategic interests in the game (see Harsanyi 1977d).

3   *The tracing procedure: a Bayesian approach to defining solutions for noncooperative games.* Another application of Bayesian ideas in game theory has been in the *tracing procedure*, which is an attempt to extend the Bayesian approach from one-person decision problems to the n-person decision problem of finding a solution for an n-person noncooperative game (Harsanyi 1975).

Most game theorists agree that the solution of any noncooperative game must be an *equilibrium point*.[3] This is so because any reasonable solution must be a strategy combination $s = (s_1, \ldots, s_n)$ with the property that no player i will have a clear incentive to *deviate* from his solution strategy s_i merely because he expects the other players to *follow* their own solution strategies $s_1, \ldots, s_{i-1}, s_{i+1}, \ldots, s_n$. But if any given player i's solution strategy s_i were *not* a best reply to the other players' solution strategies then he *would* have such an incentive. Consequently, *every* player's solution strategy *must* be in fact a best reply to the other players'. But this means that the solution must be an equilibrium point.

Nash (1951) has shown that every finite game (every game with a finite number of players, and with a finite number of pure strategies for each player) has at least one equilibrium point. But, unfortunately, equilibrium points are very far from being unique: almost every interesting game has more than one equilibrium point; indeed, it will often have a very large number, and may even have infinitely many. The purpose of the tracing procedure is to choose *one* specific equilibrium point of any noncooperative game as the solution of this game.

Let me now introduce some notations. Suppose that G is an n-person noncooperative game in which player i ($i = 1, \ldots, n$) has K_i different pure strategies, to be denoted as $a_{i1}, a_{i2}, \ldots, a_{iK_i}$. A mixed strategy s_i of player i will be a probability vector of the form $s_i = (s_{i1}, s_{i2}, \ldots, s_{iK_i})$, where $s_{ik}(k = 1, 2, \ldots, K_i)$ is the probability that this mixed strategy s_i assigns to the pure strategy a_{ik}. A strategy combination of the n players will be written as $s = (s_1, \ldots, s_n)$. A strategy combination of the $(n-1)$ players *other* than player i will be written as $\bar{s}_i = (s_1, \ldots, s_{i-1}, s_{i+1}, \ldots, s_n)$. We will also write $s = (s_i, \bar{s}_i)$. Player i's payoff function will be written as $U_i(s_i, \bar{s}_i)$.

The tracing procedure is based on the following model. Any player $j \neq i$ will have to form expectations about the strategy that

---

[3] Let me remind the reader of the relevant definitions. A given strategy s_i of player i is a *best reply* to the strategies $s_1, \ldots, s_{i-1}, s_{i+1}, \ldots, s_n$ used by the other $(n-1)$ players if s_i maximizes player i's payoff $U_i(s_1, \ldots, s_{i-1}, s_i, s_{i+1}, \ldots, s_n)$ when all other players' strategies are kept constant. A given strategy combination $s = (s_1, \ldots, s_n)$ is called an *equilibrium point* if the strategy s_i of *every* player i is a best reply to the strategies $s_1, \ldots, s_{i-1}, s_{i+1}, \ldots, s_n$ of the other players.

player i will use in the game. In accordance with Bayesian theory, it is assumed that these expectations will take the form of a *subjective* probability distribution $p_i = (p_{i1}, p_{i2}, \ldots, p_{iK_i})$ over player i's pure strategies, where p_{ik} ($k = 1, 2, \ldots, K_i$) is the subjective probability that player j assigns to the hypothesis that player i will use his pure strategy a_{ik}.

It is assumed that *any* player $j \neq i$ will derive his subjective probability distribution p_i over i's pure strategies from the basic parameters of the game (*e.g.* from the payoff functions) by the *same* computation procedure, so that different players j will arrive at the *same* distribution p_i. This distribution p_i will be called the *prior probability distribution* over player i's pure strategies.

Mathematically, such a prior distribution p_i, being a probability distribution over i's pure strategies, will have the nature of a *mixed strategy* by player i. But of course it will have a very different intuitive interpretation from an ordinary mixed strategy. Whereas an ordinary mixed strategy s_i always involves the assumption that player i *himself* will randomize among his alternative pure strategies, a prior probability distribution p_i does not involve any such assumption. Rather, it merely expresses an *uncertainty* in the *other* players' minds about the actual strategy that player i will use in the game.

I will write $p = (p_1, \ldots, p_n)$, $\overline{p_i} = (p_1, \ldots, p_{i-1}, p_{i+1}, \ldots, p_n)$, and $p = (p_i, \overline{p_i})$. Thus, $\overline{p_i}$ is a probability vector summarizing player i's expectations about the behaviour of the *other* $(n-1)$ players' behaviour in the game, whereas p summarizes all n players' expectations about one another's behaviour.

Suppose we know the vector p. How can we then predict the outcome of the game? The simplest answer would be this. Since the expectations of each player i about his fellow players' behaviour are expressed by subvector $\overline{p_i}$ of vector p, he will use that particular strategy s_i^0 that is his *best reply* to $\overline{p_i}$, *i.e.* he will use that particular strategy $s_i = s_i^0$ that maximizes his expected payoff, defined by the expression

(3.1) $U_i(s_i, \overline{p_i})$.

Accordingly, one may try to argue that the outcome (solution) of game G should be defined as the strategy combination $s^0 = (s_1^0, \ldots, s_i^0, \ldots, s_n^0)$, consisting of these best-reply strategies $s_1^0, \ldots, s_i^0, \ldots, s_n^0$. I will describe this view as the *naive Bayesian approach*.

Regrettably, this simple-minded naive Bayesian approach is

unacceptable because the strategy combination s^0 in general *will not be an equilibrium point* of game G.

Yet, even though this naive Bayesian approach would be an unsatisfactory solution theory, it can be used as a starting point for a more satisfactory theory. In fact, the tracing procedure always starts with this best-reply strategy combination s^0, but then systematically and continuously transforms this strategy combination until in the end it *converges* to an equilibrium point $s^* = (s_1^*, \ldots, s_n^*)$ of the game, which then will be chosen as the solution. Indeed, the tracing procedure is meant to model the process of *convergent expectations* by which rational players, starting from a state of expectational disequilibrium, can make their expectations converge upon a specific equilibrium point of the game.

The tracing procedure is based on a one-parameter family of auxiliary games G^t, with $0 \leq t \leq 1$. In any given game G^t, the payoff function U_i^t of each player i will be defined as

(3.2) $\quad U_i^t(s_i, \bar{s}_i) = (1-t)U_i(s_i, \bar{p}_i) + tU_i(s_i, \bar{s}_i)$.

Clearly, for $t = 1$, game $G^t = G^1$ will have the same payoff functions

(3.3) $\quad U_i^1(s_i, \bar{s}_i) = U_i(s_i, \bar{s}_i)$

as the original game G has. Therefore, we can write $G^1 = G$. On the other hand, for $t = 0$, game $G^t = G^0$ will have payoff functions of form (3.1) since

(3.4) $\quad U_i^0(s_i, \bar{s}_i) = U_i(s_i, \bar{p}_i)$

Equation (3.4) shows that game G^0 is a game of very special form which makes each player's payoff depend only on his *own* strategy s_i, and makes it independent of the other players' strategy combination s_i. Consequently, G^0 can be factored into n independent maximization problems, one for each player, since each player's task in the game will be simply to maximize the expression $U_i(s_i, \bar{p}_i)$, regardless of what the other players are doing.

For each game G^t, let E^t be the set of all equilibrium points of G^t. By Nash's theorem (Nash 1951), each of these sets E^t will be nonempty. Let P be the graph of the correspondence $t \rightarrow E^t$, $0 \leq t \leq 1$. For any fixed t, this graph will have as many points as the number of equilibrium points in E^t. Accordingly, in nondegenerate cases, in the region $t = 0$, this graph P will have only *one* point x^0, because each player i will have a *unique* (pure) strategy $s_i = s_i^0$ maximizing the payoff function $U_i^0 = U_i(s_i, \bar{p}_i)$, so that the equilibrium-point set E^0 will contain only the *one* point $s^0 = (s_1^0, \ldots, s_n^0)$. Moreover, one can show that, in nondegenerate cases, graph P will contain exactly *one* continuous path L, leading from this one point x^0 in region $t = 0$ to some point x^1 in region $t = 1$. This point x^1 will always correspond to a specific equilibrium point $s^* = (s_1^*, \ldots, s_n^*)$ of the original game $G = G^1$, and this equilibrium point s^* will be chosen as the *solution* of game G.

The mathematical procedure selecting s^* as the solution is called the *tracing procedure* because it is based on tracing (*i.e.* following) this continuous path L from its starting point x^0 (corresponding to the strategy combination s^0) to its end point x^1 (corresponding to the solution s^*).

This version of the tracing procedure, called the *linear* tracing procedure, will always yield a unique solution s^* in all nondegenerate cases. But in degenerate cases it may yield several (or even infinitely many) solutions because

(i) Graph P may contain more than one point x^0 in region $t = 0$ (in which case it will always contain infinitely many points x^0).

(ii) Moreover, even if x^0 is unique, there may be several (or even infinitely many) continuous paths leading from point x^0 to points in region $t = 1$.

It can be shown, however, that a modified form of the procedure, called the *logarithmic* tracing procedure, will *always* yield a unique solution, even in highly degenerate cases.

The solution defined by the tracing procedure always has the form $s^* = T(G,p)$. In other words, the solution s^* defined by the tracing procedure will not only depend on game G for which the solution is defined, but will also depend, and often quite sensitively depend, on the vector $p = (p_1, \ldots, p_n)$ of prior probability distributions chosen as a starting point. This of course means that the tracing procedure in itself does not provide a complete solution theory for noncooperative games. It yields a complete solution theory only if it is supplemented by a theory of how to choose suitable *prior probability distributions* p_i for any given game G. But, in conjunction with such a theory, it does represent a very attractive approach to defining solutions for noncooperative games. (For a more detailed discussion of the tracing procedure, and for the proofs of the relevant mathematical theorems, see Harsanyi 1975. For a discussion of the prior distributions see Harsanyi 1977c. The last-mentioned paper also contains a general discussion of the Harsanyi–Selten solution theory, based on the tracing procedure, and on suitable procedures for constructing prior probability distributions. See also Harsanyi 1977e.)

4 *Conclusion.* I have tried to show on two examples that, even though individual decision theory and game theory are dealing with very different classes of decision problems, a use of Bayesian decision-theoretical tools in game theory can produce valuable results.

University of California, Berkeley
University of Bielefeld

REFERENCES

Debreu, Gerard. 1959. *Theory of Value*. New York.

Harsanyi, John C. 1953. Cardinal utility in welfare economics and in the theory of risk-taking. *Journal of Political Economy* **61**, 434–5. Reprinted in Harsanyi 1976: 3–5.

Harsanyi, John C. 1955. Cardinal welfare, individualistic ethics, and inter-personal comparisons of utility. *Journal of Political Economy* **63**, 309–21. Reprinted in Harsanyi 1976: 6–23.

Harsanyi, John C. 1967–8. Games with incomplete information played by 'Bayesian' players. Parts I–III. *Management Science* **14**, 159–82, 320–34, and 486–502.

Harsanyi, John C. 1975. The tracing procedure. *International Journal of Game Theory* **4**, 61–94.

Harsanyi, John C. 1976. *Essays on Ethics, Social Behavior, and Scientific Explanation*. Dordrecht.

Harsanyi, John C. 1977a. *Rational Behavior and Bargaining Equilibrium in Games and in Social Situations*. Cambridge.

Harsanyi, John C. 1977b. Morality and the theory of rational behavior. *Social Research* **44**, 623–56.

Harsanyi, John C. 1977c. A solution concept for n-person noncooperative games. *International Journal of Game Theory* **5**, 211–25.

Harsanyi, John C. 1977d. Time and the flow of information in non-cooperative games. In *Quantitative Wirtschaftforschung*, ed. H. Albach *et al.*, pp. 255–67. Tübingen.

Harsanyi, John C. 1977e. A solution theory for noncooperative games and its implications for cooperative games. *Working Paper CP–401 Center for Research in Management Science*. Berkeley.

Harsanyi, John C. 1978. Bayesian decision theory and utilitarian ethics. *American Economic Review Papers and Proceedings* **68**, 223–8.

Nash, John F. 1951. Noncooperative games. *Annals of Mathematics* **54**, 286–95.

von Neumann, John & Morgenstern, Oskar. 1947. *Theory of Games and Economic Behavior*, 2nd ed., Princeton.

12 *Welfare, rights and fairness*

FREDERIC SCHICK

1 Most of the papers in this volume comment on Braithwaite's views on science. This is only right, for Braithwaite's analysis of the logic of science is his major contribution. But there is also a short book by him on very different matters. In 1954, he was appointed Knightbridge Professor of Moral Philosophy at Cambridge. He took his new title to heart. *Theory of Games as a Tool for the Moral Philosopher* was his inaugural lecture.

This was totally unlike any other ethics of that time. Its very title contrasts sharply with those of the books then most read, *Ethics and Language*, *The Language of Morals*, *The Place of Reason in Ethics*. The current analyses were meta–ethical. Their project was to study the nature of moral evaluation and argument. Braithwaite took a more practical line. He held that ethics had to show how conflicts should be resolved, that the philosopher's proper concern was with real-life problems. He discusses the case of a pianist and a trumpeter who live on opposites sides of a wall that doesn't block any sound. Each is bothered by the other's playing, but both are not bothered to the same degree. The utilities that each sets on his own playing alone, on the other's playing alone, on both of them playing and on neither of them playing are assumed to be cardinally measurable. How should the moral judge decide how long they each should play?

Braithwaite's attention to practical issues was a decade ahead of its time. The turmoil of the 1960s shook up philosophy along with all else. It led to students of ethics debating abortion, civil disobedience, war and peace. The temper was more that of advocacy than of cool mediation, but the focus was on the realities. The questions dealt with issues of justice: which of several opposing claims had the strongest case? Who had a right to what? What did fairness require and what did it rule out?

These were very like Braithwaite's questions. Braithwaite too saw the ethicist's problems as that of defining equity. In this, his book stands out again, here from a mode of thinking he himself endorses. He notes that his work is in the tradition of welfare economics. His technical apparatus is certainly welfarist, but the issues he raises are not. They stem from a separate tradition usually labelled *liberal*.

More on Braithwaite's position later. Let me here say only that Braithwaite does not draw the distinction I am making. He has no need to draw it, for it doesn't affect his results. There are, however, contexts in which neglecting it makes for trouble. I shall consider one such case, the so-called *liberal paradox*. I will then take up a problem that lies in wait for any liberal.

2　A welfarist thinker is one who holds that the first obligation of society is to promote the general welfare. He studies how to make the arrangements to which this directs us. Contemporary welfarist writers bring the policy down to the individual, seeing the general welfare as a function of the preferences of the people involved. It does not matter what these are, or how they were established. Where a person gets his way, he is then better off. And the general welfare is a function of how well-off people are.

We can put this more clearly. Every social welfare problem has to do with some group of people and some set X of alternatives. Each person in the group is assumed to have a utility-ranking (possibly only ordinally significant) of the items of X. A welfarist holds that the social decision must be made from the *choice-set* of X: the set of those items of X no one of which has a welfare-rank lower than some other item of X. The general welfarist problem is to specify a rule defining a welfare-ranking in every context of utility-rankings of the items of X, whatever X might be, a rule that looks only to the utility-rankings involved in each case. A rule of this sort is called a *social welfare function*, or *SWF*.

Not every SWF will be acceptable to everyone. This takes us back to meta-ethics. What conditions must any proper SWF meet? Or better, what conditions do differently oriented thinkers impose? In a well-known paper, Sen (1970a) considers the position of the liberal.[1] He suggests that the minimal liberal requirement is that an SWF square with the following: there are two (or more) people each of whom is decisive over some pairs of items in some (the same) X. On his reading, a person i is decisive over a pair-subset $\langle p,q \rangle$ of X where, if i ranks p above q utility-wise, p is set above q in the welfare-ranking of X. Sen labels this very weak principle *minimal liberalism*, or L^*. He goes on to note that most authors require that an SWF satisfy the *Pareto condition*, P: if every person utility-ranks some item of X above a second, the welfare-ranking of X sets the first above the second. Sen then shows that no SWF can always satisfy both P and L^*.

The proof is simple. We need only to find one case in which both can't be satisfied. Suppose that i is decisive over $\langle p,q \rangle$ and i' over $\langle r,s \rangle$, both pairs subsets of X. Let i rank p above q, let i' rank r above s, and let everyone (including i and i') rank q above r and s above p. Then, by L^*, the welfare-ranking of X must set p above q; by P, it must set q over r; by L^*, it must set r over s, and by P again, s over p. This would make for a cycle, and cycles are ruled out by the fact that welfare-rankings are transitive throughout (in order to generate a choice-set however X is cut down).[2] So no SWF can satisfy both P and L^* in this case.

For anyone who endorses both these principles, here is unsettling news. No decision procedure can respect unanimity and also those special preferences of people that pertain just to their own affairs. This has come to be known as the *liberal paradox*, and a sizable literature has grown up around it.[3] I don't see, however, why a liberal should worry, for he does not endorse L^*, or rather, he doesn't on the reading above. The formula itself is central for him, but he means something different by it. Nor will any other condition on SWFs capture the liberal's view. Sen's argument breaks down at the start, for it assumes that the liberal is a welfarist, that he thinks in terms of welfare, if in a special way.

The liberal departs from the welfarist's line at this very point: he denies that the general welfare overrides all. He need not be indifferent to considerations of welfare, but he insists that justice takes precedence. This conviction of the priority of justice he shares with

---

[1] Sen speaks of social *decision* functions rather than of social welfare functions, but this difference does not affect our analysis.

[2] The transitivity of welfare-rankings is part of the concept of SWFs; see Sen (1970: 41).

[3] This is surveyed in Sen (1976: 217–35).

many non-liberal thinkers – justice was basic even for Plato. What sets the liberal's position apart is his hinging of justice on who gets what. Each of the authors that comes to mind – Rawls and Dworkin stand out on the left and Nozick and Hayek on the right – sees the objects of justice-assessments as distributions of some sort, distributions of money, or of property and prospects. (They also attend to more comprehensive alternatives, but only in so far as these have distributional aspects.)

We need a general term for what a distribution distributes. I shall speak of *goods*, taking this in its broadest sense. Cars and kitchen equipment are goods, and so are houses and land and money. So also are influence and status and opportunity. Goods also cover the absence of certain burdens: certain duties, obligations and prohibitions. The liberal thinks in terms of justice, and this, for him, depends on people's holdings. He thinks of justice as a function of the specific goods different people have.

There are two points here. The liberal keeps his eyes on justice, not on the general welfare. He will perhaps argue that you have the right to chain-smoke to death, that it would be unjust for others to stop you. He need not be convinced that your dying would leave everyone better off. Also, he focuses on what people have, not on their feelings about this. A person's view of his situation enters at most indirectly. A well-adjusted slave approves the slave-economy. He prefers remaining a slave to being set free. A liberal's concept of justice calls for this preference being overruled.

Still, this is only a part of it. The cases of the slave and the smoker themselves suggest what is being left out. A liberal, after all, is committed to liberty. He has a special regard for freedom. His discussions often insist that serving justice is a step-wise affair.[4] First a society must secure people's *rights*. Only then may it turn to the rest, to what might be called the demands of *fairness*. This brings out why a liberal indeed needs something like L^*. He needs to affirm the existence of rights, to say that the first step (protecting liberties) is not vacuous.

Getting this two-step procedure clear calls for rethinking decisiveness. On a liberal's understanding, i is *decisive* over a pair $\langle p,q \rangle$ where p assigns more of some good to i than q does, and q is, at least

---

[4] Rawls speaks of the lexicographical ordering of his two principles (1971: 42ff). See also Nozick's comment on Sen in Nozick (1974: 164–6). Dworkin's concept of the *seriousness* of rights also comes to the same (1978).

partly for that reason, excluded from consideration for choice. Where there is some i decisive over $\langle p,q \rangle$, we shall speak of q as *inadmissible*. The set of all the members of X that are not inadmissible is its *admitted-set*. The admitted-set of X is the set of those members of X of which a fairness-ranking is to be found, and thereby also a choice-set. Notice that the choice-set here is taken not from X but from the admitted-set of it. Also that the choice-set is based not on a welfare-ranking of the items in this set but on a fairness-ranking of them. The liberal's choice-set is the set of those items of the admitted-set of X no one of which has a fairness-rank lower than some other item of this admitted-set.

The liberal does endorse L^*: there are two (or more) people each of whom is decisive over some pair of items in some (the same) X. But his concept of decisiveness renders this different from Sen's version of it. The liberal also endorses the much stronger principle L. This says that *every* person is decisive over some pair of items in some (the same) X. A liberal does not endorse P, so there is no way in which this L (or the weaker L^*) can conflict with it for him. And though P too might be reshaped for a liberal (I touch on this in the next section), the new P could not conflict with L (or L^*) either, for the two would still be kept apart by the priority of L (L^*).

The possibility of disagreement remains. The welfarist may establish a choice-set that contains an element inadmissible for the liberal, and a liberal's choice-set may contain some items the welfarist's does not. The two ideologies divide on the basics: does the general welfare come first or does justice come first? Must the service of justice promote welfare or must the promotion of welfare serve justice? There is a difference of perspectives here, but if we keep this in mind there are at least no paradoxes.

3   The welfarist's social welfare problem is to specify an SWF, a rule that defines a welfare-ranking of the items of any X solely in terms of the utility-rankings of the people involved. A liberal's social *justice* problem is two-fold. The liberal needs to set up a rule that identifies the admitted-set of any given X. And he must set up a second rule defining a fairness-ranking of the items of any admitted-set. Call a rule of the first sort a *rights-rule*, and a rule of the second sort a *social fairness function*, or *SFF*.

Rights-rules are familiar. Mill's "simple principle" is that 'The sole end for which mankind are warranted, individually or collectively, in interfering with the liberty of action of any of their number, is self-protection' (1961: 197). There is also Rawls' "first principle": 'each person is to have an equal right to the most extensive basic liberty compatible with a similar liberty for others' (1971: 60). Nozick's entitlement principles (1974: 150ff) spell out a third idea of rights. On Nozick's view, the theory of rights indeed exhausts the theory of justice. (Conservative and left-liberals differ mainly in this, that the former make admitted-sets singletons, or close to it, while the latter leave room for fairness.)

Rights-rules need not look the part. Some of the to-each-according-to-m dicta hint at people's rights – for instance, to each according to his *needs*. (The others are best read as fairness functions.) Consider also the American Bill of Rights: 'Congress shall make no law respecting the establishment of religion, or prohibiting the free exercise thereof; or abridging the freedom of speech, or of the press; or . . .' In our terms, this says that requiring some particular religious observance is not part of any admissible course, nor is forbidding any form of worship. . .

The metastudy of rights-rules has a respectable history, but there is little to show for it. (Excepting L and L^*, which only say that there are rights, not anything *about* them.) The trouble is, there is no agreement on what a person's rights depend upon, no accepted view that rights are functions of this or that variable or family of variables. The discussion is still at the stage of considering different possibilities – that people's rights depend on the claims they are entitled to make, or on the obligations of others, *etc*.

Let us turn to the problem of fairness. Many positions have been developed; maximizing, maximining and egalitarian SFFs all have their supporters, and there are still other kinds.[5] The metastudy of SFFs is a totally unexplored field. But here an opening can be seen. Each of the principles of social welfare has a fairness analogue. All we need do to change one into the other is to replace the mention of utility functions with a mention of some appropriate functions of goods and to refer to fairness in place of welfare. This means that the various impossibility results of welfare theory carry over, and also the proposals for getting around them. Whether the fairness counterparts of the welfare principles retain their force is another matter. Certainly, some will be challenged. (Take the rephrased Pareto principle. Is distribution d always *fairer* than d' where

[5] I discuss SFFs of various sorts in Schick (1980a).

everyone has more goods in d than in d'? What if d is egalitarian and there are vast disparities in d'?)

There is a prior problem with the SFFs themselves. An SFF rank-orders distributions in the light of the goods these assign to each person. We must here distinguish: fairness is not a function of the mere bulk of these goods, but of their potential contribution to the projects of the people involved. I shall speak of the *resource*-value of a person's goods. Two books are twice the goods of one, but a second copy of a book I already have adds very little to my resources. If you have twice my money, twice my land, twice my credit, *etc.*, you have twice my goods. It does not follow that you have twice my resources. An SFF ranks distributions in terms of the total resources they assign, or in terms of the resources of the person who has least, or of their departure from the equal-resources distribution, or of some other resource-based variable. Some way of measuring resources is always presupposed.

What sort of measurement must this be? Not every SFF requires resources to be cardinally measurable. The maximizing SFF calls for this, the maximining SFF does not. All, however, require some measurement of resources that allows for the interpersonal comparison of them. They presuppose some basis for saying that a person i has more resources under distribution d than i' has, or that i has more under d than i' has under d'. The problem here resembles the problem of the cardinality and comparability of utilities. I shall discuss two approaches to it, one worked out by Coleman, the other adapted from an idea of Arrow's.

Coleman (1972, 1973) assumes the existence of a market in which all goods are freely traded. On his view, goods are held in the form of control over issues, either partial or complete control. People trade the control they have over issues in which they are little interested for more control over issues in which their interest is greater. The layout of the control people have over issues and of their interest in them establishes each person's resources.

The central concepts here are those of *issues*, *interest in* them and *control over* them. An *issue* is any pair of contingencies of the form *h will happen, h will not*; or rather, any one of a comprehensive set of logically independent such pairs – independent in the sense that no outcome of one implies either outcome of any of the others. A person's *interest* in an issue is the difference between the utilities he assigns to its two outcomes divided by the sum of the corresponding differences for all the issues involved. The *control* a person has over an issue is the influence he has on how that issue comes out. Let *o* be one of the outcomes (either one) of some issue *j*, and let some person be in a position either to do *a* or to do *a'* – call these two of his *options*. This person has *k* control over *j* where the difference between the conditional probabilities of *o* on his doing *a* and on his doing *a'* is *k*, and (further) he has no other pair of options such that the difference between the corresponding conditional probabilities is greater than *k*.

One more technical concept, that of the *market value* of (full) control over an issue. This is the weighted sum of every person's interest in this issue, the weights being each person's resources. Let $v_j(d)$ be the market value of control over *j* where the distribution (of control) is *d*. Let $c_{ij}(d)$ be the control over *j* that *d* assigns to *i*. Coleman's view is that *i*'s *resources* under *d* are

(1) $r_i(d) = \sum_j v_j(d)c_{ij}(d)$

That is, *i*'s resources under *d* are the weighted average of the control *d* assigns him over the various issues, the weights being the market values of control over these issues under *d*.

Given some simplifying assumptions and sufficient information about every person's interests, it is possible to solve for all the $v(d)$s, and so to establish each person's resources under any distribution. It turns out that the sum of all resources must be 1. A person's $r(d)$ is thus an index of the proportion of the total resources that are his, and so resources are interpersonally comparable.

Coleman defines resources in terms of market values, and so in terms of interests and thus of utilities. Does this blur the line we have drawn between the position of the liberal and the welfarist? No, for (1) does not identify resources with utilities: a person's resources under *d* are not the utility he sets on *d*'s holding. The analysis indeed erases distinctions we should be keeping clear. Since the sum of everyone's resources always is the same, every distribution is simply egalitarian. Maximizing is rendered trivial – it always sets all alternatives on a par – and maximining loses its character. But a simple extension of the theory allows for getting around this.[6]

There is, however, a problem here I know of no way of handling. A person's control over an issue is defined in terms of that pair of his options that gives him the biggest impact on how the issue comes out. Suppose that, for some issue, a person's most effective

---

[6] These matters are discussed in Schick (1980a).

option-pair is his making or not making a fool of himself, or perhaps his committing or not committing some crime. He can exercise the influence he has only at some cost. The prospect of paying the price will certainly temper his willingness to act. Shouldn't this affect our assessment of the control this person has? (Suppose I can make something happen by pressing a button, and can make sure it doesn't by pressing another. You too can make something happen by pressing a button, but can only prevent it by shooting yourself. Is our control in these cases the same?)

Or take it one step further. Our willingness to exercise the control we have (or might have) must affect our readiness to yield (or to acquire) that control. Shouldn't then our willingness to act be reflected in the market values of control? Coleman's answer is *no* – it is also *no* to our question concerning the assessment of control itself – but he hasn't given any rationale for this. So his analysis does not provide a fully general theory.[7]

Let me turn to a second approach, this one stemming from a discussion of Arrow's to a somewhat different purpose. What sense can we give to the interpersonal comparison of *utilities*? There is no problem with thinking that a person can rank-order total bundles of goods in terms of how much he himself would like to have them. This remains true if we extend the goods-concept beyond the scope we gave it above, if we have it cover not only all sorts of commodities and the rest, but also those special characteristics of people that affect their reactions to having various bundles. Let the (extended) bundles of goods thus include all these predisposing traits – wealth, colour, sex, *etc.* – whether they stem from people's genes or from their distinctive positions in life. One of these bundles is the one that i now has, another is the bundle he would have under a different (extended) distribution d', a third is the bundle that i' has under the current distribution d, a fourth the bundle that i' would have under d', *etc.* The supposition that i can rank-order all these bundles is Arrow's thesis of *extended sympathy* (1963, 1977).

Arrow is content to speak of the ordinal ranking of these bundles. It is, however, generally accepted that if a ranking is inclusive enough, and if it meets certain conditions, it makes for cardinal measurability – that is, it determines a utility function unique up to a linear transformation. (This is generally accepted, but not universally, Arrow himself being one of the hold-outs.) I shall thus take

[7] There are still further difficulties; these are discussed in Schick (1980).

the thesis to be that each person has a utility function (unique up to a linear transformation) defined over the set of all the (extended) goods bundles.

One further refinement. The bundle of goods that d assigns i will be taken to be what he has (or will have) when all exchanges are done. This assumes that the market is determinate, that it always specifies a single bundle for each person. Less would do for us too: if someone might wind up with any one of several bundles, it makes no difference to him which it is. In this case, "his bundle" is any one of these several. Another way of putting it is that, from here on (unlike with Coleman), we keep to post-exchange valuations.[8]

Let $G_i(d)$ now be the property of having the bundle of goods that distribution d assigns to i. Let $iG_i(d)$ be the proposition that i has this property, $i'G_i(d)$ the proposition that i' has it, *etc.* Arrow holds that every person ranks the various bundles the same way. On our cardinal reading of this, the idea is that

(2) $u_{i'}(i'G_i(d)) = u_{i''}(i''G_i(d))$

for every i, i', i'' and d, the $u(.)$s being cardinal utility functions. In simple prose, the utility that a person i' sets on his having the bundle that d assigns to i is the same as the utility that any other person i'' sets on his (i'') having that bundle. Arrow goes on to suggest that

(3) $u_i(d) = u_i(iG_i(d))$

The utility that i sets on a distribution is the utility he sets on his having the bundle this distribution assigns to him.

Equation (2) asserts the interpersonal comparability of utilities. It refers to propositions of the $i'G_i(d)$-sort only, but it suffices to establish comparability throughout. (Notice that it confines us to coordinated transformations: if the origin and unit of one person's scale is changed in a certain way, the origin and unit of all other people's scales must be changed the same way.) The addition of (3) is nonetheless useful, for (2) and (3) together imply that a person need only know how he values different bundles in order to compare people's reactions to different (or the same) distributions. Suppose that $u_{i'}(i''G_i(d)) > u_{i'}(i''G_i(d'))$ and that i'' knows that this holds. Then he is in a position to say that $u_i(d) > u_{i'}(d')$.

But (3) is clearly implausible. Not every person reacts to distributions solely on the basis of what he gets out of them. Most people are to some extent altruist (and malicious), reacting also to what

---

[8] If we had kept to these above, we could have simplified (1) by putting i's interest in j in place of the market value of j. (In this simpler (1), c_{ij} would be i's *post-exchange* control over j.)

others get. Where a person has some ideal of fairness, the pattern of the allotment may count for him too. A return to our proper business gets us clear of these matters. None of them holds where we speak of resources rather than of utilities. So let us replace (3) with

(4) $\quad r_i(d) = u_i(iG_i(d))$

The *resources* of i under d are now measured by the utility i sets on his having the bundle that d assigns him. By (2), this is the same as the utility all others set on their having this bundle. On the analogy of the above, (2) and (4) together allow for interpersonal *resource-comparisons*.

It may be objected that the distributions we have brought in cover more than society can manage. Tastes and talents and other biasing traits are dispensed by nature alone. There is no way for these to be traded or shifted from one person to another. But this only says that the set Y of all possible distributions is larger than the set X of alternatives of any social justice problem, that any X is bound to be a proper subset of Y. An X can include only those items of Y that assign the non-tradable goods as these are in fact already assigned. All the alternatives in any X must already hold in their non-tradable parts. (This requires expanding an assumption with which we started: each person must have a utility-ranking of the items of Y, not just of X.)

Another objection to (4) might be that it merges resources into utilities. We noted a similar charge against (1). Here, as there, the distinction remains: (4) defines i's resources under d independently of the utility i sets on d. It refers to the utility i sets on his having the bundle that d assigns him. The utility i sets on d itself may be either higher or lower.

There is, however, a more serious problem deriving from the idea expressed in (2). This idea is very old, going back to Rousseau, whose theory of the social contract is based on the supposition that if people gave up their goods, their interests on most issues would coincide. Their *individual wills* differ, but this is because their different possessions bias them in different ways. If they no longer had any possessions, their biases would dissolve. What would then come out is their deeper *general* wills, and the general will is the same in all. Not only are their desires and preferences the same on this deeper level, but also the intensities of feeling behind them. Equation (2) might be called *Rousseau's principle* (1950, 1950a).

The question is, what are people like when the general will speaks through them? Or simply, how are we to think of i' and i'' in (2)? The purpose of (2) requires that they have no traits that might ever lead them to value any single bundle differently. They can't be either rich or poor, clever or stupid, good-looking or plain, black or white or yellow or brown, young or old, male or female. But a creature without age and sex is not any sort of person at all. Trying to make (2) plausible only makes it incomprehensible.

This undoes our rationale for interpersonal comparisons. On our analysis, for i'' to be entitled to compare i's resources with those of someone else, it must be true (*inter alia*) that $u_i \cdot \cdot (i''G_i(d)) = u_i(iG_i(d))$. We can be sure this equation holds only where both i'' and i are mere spectres. The party making the comparison must be bloodless, and also those others whose resources are compared. Interpersonal comparisons call for depersonalizing people. What we have endorsed is only inter*spectral* comparisons.

A closer look at Rousseau may help us to get further. The Rousseauian rationale does not require that people actually give up their goods, but only that they think they give them up, or no longer think they have them. The core requirement is that people do not know what goods they have. Let $u^*(.)$ be the utility function a person would have if he did not know which bundle the current distribution assigned to him. We can replace (2) with

(5) $u^*_i \cdot (i'G_i(d)) = u^*_i \cdot \cdot (i''G_i(d))$

and go on to replace (4) with

(6) $r_i(d) = u^*_i(iG_i(d))$

A person's resources under d are here measured by the utility he *would* set on his having the bundle that d assigns him *if* he did not know which bundle was currently his. There is no longer a problem with the corporeality of the parties. The is can have all the traits of full-bodied people.

But (5) makes for new uneasiness. It draws on credit it may not have. The assumption of the existence of u-functions unique up to a linear transformation does not suffice to establish that each person also has such a u^*-function. The logic of evaluation behind the veil of ignorance (where no one knows what goods are his) has been much studied lately, but the proposed analyses don't help here. Rawls (1971) suggests that, where people are rational, they will, behind the veil, set on every distribution the utility that the person who gets the least in it sets on his having the bundle that this distribution assigns to him. This does not bear on people's valuations of propositions of the $i'G_i(d)$-sort, but on their valuations of distributions *in toto*. It gives us (where people are Rawlsian-rational) not (5) but

(7) $u^*_i\cdot(d) = u^*_i\cdot\cdot(d)$

Here we now are left stranded. No analysis of resources is in sight. We clearly cannot accept

(8) $r_i(d) = u^*_i(d)$

For this (with (7)) implies that each person's resources are always the same as anyone else's. In the context of (8), all talk of fairness is without any point. It makes no difference what we arrange, for every distribution is necessarily egalitarian.

A second suggestion is that of Harsanyi (1977, 1977a), who holds that, behind the veil of ignorance, a rational person sets on each distribution the average of the utilities that all those involved set on their having the bundles that the distribution assigns them. This again gives us not (5) but (7), and so again leaves us without any way of going on to a definition of resources.[9]

We must retreat and start again, but it isn't obvious where. What is obvious is only that the problem of measuring resources remains open. The foundations of the theory of fairness are yet to be laid.

4 Our course has taken us far from Braithwaite and his two musicians. What with spectres and happy slaves, it may at times have seemed other-worldly. I believe that the framework drawn provides for the earthbound liberal's view. But I concede that the liberal proper is an uncommon species. There are more quasi- and near-liberals than all-the-way liberals.

Take Braithwaite's case of the pianist and the trumpeter. The good to be alloted is time, playing-time and silence-time. The distributions at issue are the conjunctions of the possible time-allotments to each with the current assignment of all other goods. What is striking here is this: Braithwaite does not consider how the musicians fare in the fixed part of the distributions. The pianist's being rich and the trumpeter poor does not affect the judgement of fairness. All that matters is what these two get out of each time-division. This might be called *restricted*-focus liberalism, liberalism proper being *full*-focused. For a restricted-focus liberal, the problem of the measurement of resources is different from that above.

[9] This would not disturb Harsanyi, for he is a welfarist, not a liberal.

People's total resources under distributions need no longer be known, but only the resource-contribution of their share of the goods now allottable.

We might hope to hold things together by trimming at an earlier point. Let us take the allotments themselves to be the distributions at issue. Braithwaite then departs in a second way from liberalism proper. The liberal distinguishes people's resources under a distribution from the utilities that they set on it. The utilities people set are often shaped for them by those in control, so our heeding utilities could give the *status quo* an edge. (Respecting the brainwashed slave's contentment would favour the retention of slavery.) Braithwaite sees no danger of this in his particular case. He measures his musicians' gain from the allotments by the utilities that they set on them. This might be called *utility-oriented* liberalism, liberalism proper being *resource*-oriented. The utility-oriented liberal has no problem with resource-measurement, or rather, for him, this problem reduces to that of measuring utilities. It does not follow that his fairness-ranking is a welfare-ranking: a utility-oriented liberal remains committed to fairness, and it may be that an acceptable SWF is not an acceptable SFF for him.

There are also thinkers who do not make this distinction. They admit the priority of rights, but leave the items in the admitted-set to be ranked in terms of welfare.[10] Their position is expressed in the juncture of a rights-rule with an SWF. This is no longer liberalism, for it has dropped the concern with fairness. But since it still insists on rights, it is not yet welfarism. It might be called *libfarism*.

No doubt the liberal line is too exacting and narrow. Our need to decide on meagre information makes for less stringent views. And the attraction of other ideologies (welfarism among them) suggests that we compromise somehow. Still, the liberal's theory retains a special interest. It is the ideology of distributive justice, unqualified and uncompromised. Perhaps we can't go this road all the way. But we want to know where it leads, and what it would take to get there.

Rutgers University

---

[10] This is Dworkin's position in Dworkin (1978a).

REFERENCES

Arrow, K. J. 1963. *Social Choice and Individual Values*, 2nd ed., New York.

Arrow, K. J. 1977. Extended Sympathy and the Possibility of Social Choice, *American Economic Review* **67**, 219–25.

Braithwaite, R. B. 1963. *Theory of Games as a Tool for the Moral Philosopher*, Cambridge.

Coleman, J. S. 1972. Systems of Social Exchange, *Journal of Mathematical Sociology* **2**, 145–63.

Coleman, J. S. 1973. *The Mathematics of Collective Action*, Chicago.

Dworkin, R. 1978. *Taking Rights Seriously*, Cambridge, Mass.

Dworkin, R. 1978a. Taking Rights Seriously. In Dworkin (1978).

Harsanyi, J. C. 1977. Morality and the Theory of Rational Behavior, *Social Research* **44**, 623–56.

Harsanyi, J. C. 1977a. *Rational behavior and Bargaining Equilibrium in Games and Social Situations*, Cambridge.

Mill, J. S. 1961. *On Liberty*. In *The Philosophy of John Stuart Mill*, ed. M. Cohen, New York.

Nozick, R. 1974. *Anarchy, State, and Utopia*, New York.

Rawls, J. 1971. *A Theory of Justice*, Cambridge, Mass.

Rousseau, J. J. 1950. *A Discourse on the Origin of Inequality*. In J. J. Rousseau, *The Social Contract and Discourses*, New York.

Rousseau, J. J. 1950a. *The Social Contract*. In J. J. Rousseau, *The Social Contract and Discourses*, New York.

Schick, F. 1980. Some Notes on Exchange and Control, forthcoming.

Schick, F. 1980a. Toward a Logic of Liberalism, *Journal of Philosophy*, forthcoming.

Sen, A. K. 1970. *Collective Choice and Social Welfare*, San Francisco.

Sen, A. K. 1970a. The Impossibility of a Paretian Liberal, *Journal of Political Economy* **78**, 152–7.

Sen, A. K. 1976. Liberty, Unanimity and Rights, *Economica* **43**, 217–35.

Bibliography of the philosophical writings of R. B. Braithwaite

1924

Review of Ducasse, C. J. 1924. *Causation and the Types of Necessity*. Seattle.
   Mind **33**, 460–1.

1925

Critical notice of Nicod, J. 1924. *Le Problème Logique de l'Induction*. Paris.
   Mind **34**, 483–91.

1926

Universals and the 'method of analysis'.
   Aristotelian Society Supplementary Volume **6**, 27–38.
[Third contribution to symposium with H. W. B. Joseph and F. P. Ramsey
at Cambridge, July 1926.]

Critical notice of Whitehead, A. N. 1926. *Science and the Modern World*.
Cambridge.
   Mind **35**, 489–500.

1927

The State of Religious Belief: an inquiry based on 'The Nation and Atheneum'
questionnaire, 77pp. London.

Belief in a life force.
   The Nation and Atheneum 5 March 1927, 756–7.

Is the 'fallacy of simple location' a fallacy?
   Aristotelian Society Supplementary Volume **7**, 224–36.
[Second contribution to symposium with L. S. Stebbing and D. M.
Wrinch at Bedford College London, July 1927.]

'The idea of necessary connexion' I.
  Mind **36**, 467–77.

1928

'The idea of necessary connexion' II.
  Mind **37**, 62–72.

Verbal ambiguity and philosophical analysis.
  Proceedings of the Aristotelian Society **28**, 135–54.
[Read 19 March 1928.]

Time and change.
  Aristotelian Society Supplementary Volume **8**, 162–74.
[Second contribution to symposium with J. Macmurray and C. D. Broad
at Bristol, July 1928.]

1929

Professor Eddington's Gifford Lectures.*
  Mind **38**, 409–35.
[*Eddington, A. S. 1928. *The Nature of the Physical World*. Cambridge.]

1930

Obituary notice of F. P. Ramsey.
  Cambridge Review **51** (31 January), 216.

Review of Stout, G. F. 1929. *A Manual of Psychology*, 4th edn. London.
  Mind **39**, 507–9.

1931

Obituary notice of F. P. Ramsey
  Journal of the London Mathematical Society **6**, 75–8.

Editor's introduction to Ramsey, F. P. 1931. *The Foundations of Mathematics and other Logical Essays*, ed. R. B. Braithwaite, pp. ix–xiv. London.

Obituary notice of W. E. Johnson.
  Cambridge Review **52** (30 January), 220.

Indeterminacy and indeterminism.
  Aristotelian Society Supplementary Volume **10**, 183–96.
[Third contribution to symposium with C. D. Broad and A. S. Eddington
at Cambridge, July 1931.]

Critical notice of Jeffreys, H. 1931. *Scientific Inference*. Cambridge; Northrop, F. S. C. 1931. *Science and First Principles*. Cambridge; and Whyte, L. L. 1931. *Critique of Physics*. London.
> *Mind* **40**, 492–501.

1932

Lewis Carroll as logician.
> *Mathematical Gazette* **16**, 174–8.

1933

The nature of believing.
> *Proceedings of the Aristotelian Society* **33**, 129–46.
> [Read 13 February 1933.]

Imaginary objects.
> *Aristotelian Society Supplementary Volume* **12**, 44–54.
> [Second contribution to symposium with G. Ryle and G. E. Moore at Birmingham, July 1933.]

Solipsism and the 'common sense view of the world'.
> *Analysis* **1**, 13–15.

Philosophy.
> *Cambridge University Studies*, ed. H. Wright, pp. 3–32. London.

Letter to the editor.
> *Mind* **42**, 416.
> [Reply to Wittgenstein, L. 1933. Letter to the editor. *Mind* **42**, 415–16.]

1934

Review of Lenzen, V. F. 1931. *The Nature of Physical Theory*. New York.
> *Mind* **43**, 129–30.

Critical notice of Peirce, C. S. 1931–3. *Collected Papers*, vols. 1–4, ed. C. Hartshorne and P. Weiss, Cambridge, Mass.
> *Mind* **43**, 487–511.

1936

Experience and the laws of nature.
> *Actes du Congrès International de Philosophie Scientifique, Sorbonne, Paris 1935. Actualités Scientifiques et Industrielles 388–395*, pp. V.57–64. Paris.

1938

Review of Fraser, L. M. 1937. *Economic Thought and Language: a Critique of some Fundamental Economic Concepts*. London.
    Economic Journal **48**, 89–92.

Propositions about material objects.
    Proceedings of the Aristotelian Society **38**, 269–90.
[Read 21 March 1938.]

The relevance of psychology to logic.
    Aristotelian Society Supplementary Volume **17**, 19–41.
[First contribution to symposium with B. Russell and F. Waismann at Oxford, 9 July 1938.]

Two ways of definition by verification.
    Erkenntnis **7**, 281–7.

Professor Broad's mausoleum for McTaggart.*
    Cambridge Review **60** (4 November), 68.
[*Broad, C. D. 1938. *Examination of McTaggart's Philosophy*, vol. 2. Cambridge.]

1939

Review of Watson, W. H. 1938. *On Understanding Physics*. Cambridge.
    Mind **48**, 242–5.

1940

The Provost of Oriel's ethics.*
    Cambridge Review **61**, (2 February), 225.
[*Ross, W. D. 1939. *Foundations of Ethics*. Oxford.]

Critical notice of Eddington, A. S. 1939. *The Philosophy of Physical Science*. Cambridge.
    Mind **49**, 455–66.

1941

Review of Jeffreys, H. 1939. *Theory of Probability*. Oxford.
    Mind **50**, 198–200.

Review of Hardy, G. H. 1940. *A Mathematician's Apology*. Cambridge.
    Mind **50**, 420–2.

Review of Woodger, J. H. 1939. *The Technique of Theory Construction*. Chicago.
    Philosophy **16**, 419.

1942

Characterisations of finite Boolean lattices and related algebras.
Journal of the London Mathematical Society **17**, 180–92.

Review of Ayer, A. J. 1940. *The Foundations of Empirical Knowledge.*
London.
Philosophy **17**, 86–8.

1943

The new physics and metaphysical materialism.
Proceedings of the Aristotelian Society **43**, 203–9.
[Third contribution to symposium with L. S. Stebbing, J. H. Jeans and
E. T. Whittaker at the Royal Institution, London, 19 May 1943.]

1945

Review of Schilpp, P. A., ed. 1942. *The Philosophy of G. E. Moore.*
Chicago.
Philosophy **20**, 256–60.

1946

Belief and action.
Aristotelian Society Supplementary Volume **20**, 1–19.
[Inaugural Address, delivered to the Joint Session at Manchester, 5 July
1946.]

Obituary notice of J. M. Keynes.
Mind **55**, 283–4.

1947

Teleological explanation.
Proceedings of the Aristotelian Society **47**, i–xx.
[The Presidential Address, delivered 4 November 1946.]

1949

Johnson, William Ernest.
Dictionary of National Biography 1931–40, pp. 489–90. London.

1951

Moral principles and inductive policies.
Proceedings of the British Academy **36** (1950), 51–68.
[Henriette Hertz Philosophical Lecture, delivered 15 February 1950.]

1952

Reducibility.
Aristotelian Society Supplementary Volume **26**, 121–38.
[Third contribution to symposium with J. F. Thomson and G. J. Warnock at Birmingham, 13 July 1952.]

1953

Common action towards different moral ends.
Proceedings of the Aristotelian Society **53**, 29–46.
[Read 24 November 1952.]

Scientific Explanation: a study of the function of theory, probability and law in science, xii+376pp. Cambridge.
[Based upon the Tarner lectures, 1946. Reprinted with a few corrections, 1955; no further corrections in later reprints. Spanish translation: *La Explicación Científica*, tr. V. S. de Zavala. Madrid, 1965. Italian translation: *La Spiegazione Scientifica*, tr. G. Jesurum. Milano, 1966.]

The meaning of empirical probability statements.
Proceedings of the XIth International Congress of Philosophy, Brussels, **14**, 136–8. Amsterdam.

1954

The nature of theoretical concepts and the rôle of models in an advanced science.
Revue Internationale de Philosophie **8**, 34–40.

Review of Quine, W. V. O. 1953. *From a Logical Point of View*. Harvard.
Cambridge Review **75** (1 May), 417–18.

Critical notice of Hare, R. M. 1952. *The Language of Morals*. Oxford.
Mind **63**, 249–62.

1955

Theory of Games as a Tool for the Moral Philosopher, 76pp. Cambridge.
[Inaugural Lecture, delivered at Cambridge, 2 December 1954.]

Explanation.
Encyclopaedia Britannica, **8**, 971–4. London.

Report on *Analysis* problem no. 7: 'Can I decide to do something immediately without trying to do it immediately?'
Analysis **16**, 1.

A moralist's reaction to the Billy Graham films.
  Cambridge Review **77** (22 October), 64.

An Empiricist's View of the nature of Religious Belief, 35pp. Cambridge.
[The ninth A. S. Eddington Memorial Lecture, delivered at Oxford, 22
November 1955. German translation: Die Ansicht eines Empiristen über
die Natur des religiösen Glaubens, in *Sprachlogik des Glaubens*, tr, & ed.
I. U. Dalferth, pp. 167–89. München, 1974.]

1956

Religion and empiricism IV.
  Cambridge Review **77** (10 March), 433–6.
[Republished in *Christian Ethics and Contemporary Philosophy*, ed. I. T.
Ramsey, pp. 88–94. London, 1966.]

1957

On unknown probabilities.
  *Observation and Interpretation: Proceedings of the Ninth Symposium of
  the Colston Research Society held in the University of Bristol 1957*, ed.
  S. Körner, pp. 3–11. London.

Probability and induction.
  British Philosophy in the Mid-Century, ed. C. A. Mace, pp. 135–51.
  London.

1958

Review of Harrod, R. F. 1956. *Foundations of Inductive Logic*. London.
  Economic Journal **68**, 146–9.

The theory of games and its relevance to philosophy.
  Philosophy in the Mid-Century: a survey, ed. R. Klibansky, vol. 1,
  pp. 148–52. Firenze.

1959

A terminating iterative algorithm for solving certain games and related
sets of linear equations.
  Naval Research Logistics Quarterly **6**, 63–74.

Axiomatizing a scientific system by axioms in the form of identifications.
  *The Axiomatic Method: Proceedings of an international symposium held at
  the University of California, Berkeley 1957–8*, ed. L. Henkin *et al.*,
  pp. 429–42. Amsterdam.

Bertrand Russell, Philosopher.
  Nature **184**, 749–50.
[Review of B. Russell 1959. *My Philosophical Development*. London.]

1961

Review of Körner, S. 1960. *The Philosophy of Mathematics*. London.
  Cambridge Review **82** (11 March), 402.

1962

Introduction to Gödel, K. 1962. *On Formally Undecidable Propositions of*
Principia Mathematica *and Related Systems*, tr. B. Meltzer, pp. 1–32.
Edinburgh.

George Edward Moore, 1873–1958.
  Proceedings of the British Academy **47** (1961), 293–309.
[Republished with two minor corrections in G. *E. Moore: Essays in Retros-
pect*, ed. A. Ambrose *et al.*, pp. 17–33. London, 1970.]

Models in the empirical sciences.
  *Logic, Methodology and Philosophy of Science: Proceedings of the 1960
  International Congress, Stanford*, ed. E. Nagel *et al.*, pp. 224–31. Stan-
  ford.

1963

The role of value in scientific inference.
  Induction: some current issues, (proceedings of a meeting held at Wes-
  leyan University, 12–17 June 1961), ed. H. E. Kyburg Jr. & E. Nagel,
  pp. 180–93. Middletown, Connecticut.

1964

Half belief.
  Aristotelian Society Supplementary Volume **38**, 163–74.
[Second contribution to symposium with H. H. Price at Reading, 12 July
1964.]

1965

Why is it reasonable to base a betting rate upon an estimate of chance?
  *Logic, Methodology and Philosophy of Science: Proceedings of the 1964
  International Congress, Jerusalem*, ed. Y. Bar-Hillel, pp. 263–73.
  Amsterdam.